I0065006

# Sustainable Tourism X

**WIT**PRESS

WIT Press publishes leading books in Science and Technology.
Visit our website for the current list of titles.
www.witpress.com

**WIT**eLibrary

Home of the Transactions of the Wessex Institute.
Papers contained in this volume are archived in the WIT eLibrary in volume 256 of WIT
Transactions on Ecology and the Environment (ISSN 1743-3541).
The WIT eLibrary provides the international scientific community with immediate and
permanent access to individual papers presented at WIT conferences.
Visit the WIT eLibrary at www.witpress.com.

TENTH INTERNATIONAL CONFERENCE ON
SUSTAINABLE TOURISM

# Sustainable Tourism 2022

## CONFERENCE CHAIRMEN

**Stavros Syngellakis**
*Wessex Institute, UK*
*Member of WIT Board of Directors*

**Pablo Diaz Rodriguez**
*University of La Laguna, Spain*

## INTERNATIONAL SCIENTIFIC ADVISORY COMMITTEE

Alma Bojorquez-Vargas
Joao-Manuel Carvalho
Eleni Didaskalou
Mauro Dujmovic
Hiroshi Kato
Aida Macerinskiene
Sabrina Meneghello
Jose Luis Miralles i Garcia
Yasuo Ohe
Marko Peric
Lorenz Poggendorf
Manal Yehia Tawfik
Charles Tushabomwe-Kazooba

### Organised by
*Wessex Institute, UK*
*University of La Laguna, Spain*

### Sponsored by
*WIT Transactions on Ecology and the Environment*
*International Journal of Environmental Impacts*
*International Journal of Energy Production and Management*

# WIT Transactions

Wessex Institute
Ashurst Lodge, Ashurst
Southampton SO40 7AA, UK

We would like to express thanks to all the conference Chairs and members of the International Scientific Advisory Committees for their efforts during the 2022 conference season.

## Conference Chairs

**Joanna Barnes**
*University of the West of England, UK*

**Juan Casares**
*University of Santiago de Compostela, Spain*
(Member of WIT Board of Directors)

**Alexander Cheng**
*University of Mississippi, USA*
(Member of WIT Board of Directors)

**Pilar Chias**
*University of Alcala, Spain*

**Pablo Diaz Rodriguez**
*University of La Laguna, Spain*

**Andrea Fabbri**
*University of Milano-Bicocca, Italy*

**Fabio Garzia**
*University of Rome "La Sapienza", Italy*

**Massimo Guarascio**
*University of Rome "La Sapienza", Italy*

**Santiago Hernandez**
*University of A Coruna, Spain*
(Member of WIT Board of Directors)

**Massimiliano Lega**
*University of Naples Parthenope, Italy*

**Mara Lombardi**
*University of Rome "La Sapienza", Italy*

**James Longhurst**
*University of the West of England, UK*

**Elena Magaril**
*Ural Federal University, Russia*

**Stefano Mambretti**
*Polytechnic of Milan, Italy*
(Member of WIT Board of Directors)

**Jose Manuel Mera**
*Polytechnic University of Madrid, Spain*

**Jose Luis Miralles i Garcia**
*Polytechnic University of Valencia, Spain*

**Giorgio Passerini**
*Polytechnic University of Le Marche, Italy*
(Member of WIT Board of Directors)

**David Proverbs**
*University of Wolverhampton, UK*

**Elena Rada**
*Insubria University, Italy*

**Stefano Ricci**
*University of Rome, La Sapienza*

**Graham Schleyer**
*University of Liverpool, UK*

**Stavros Syngellakis**
*Wessex Institute, UK*
(Member of WIT Board of Directors)

**Borna Abramovic** University of Zagreb, Croatia

**Tawfiq Abuhantash** American University of Ras Al Khaimah, UAE

**Alejandro Acosta Collazo** Autonomous University of Aguascalientes, Mexico

**Khalid Al Saud** King Saud University, Saudi Arabia

**Ghassan Al-Dweik** Palestine Polytechnic University, Palestine

**Hind Algahtani** Imam Abdulrahman bin Faisal University, Saudi Arabia

**Abdulkader Algilani** King Abdulaziz University, Saudi Arabia

**Mir Ali** University of Illinois at Urbana-Champaign, USA

**Bakari Aliyu** Taraba State University, Nigeria

**Samar Aljahdali** University of Jeddah, Saudi Arabia

**Hussain Al-Kayiem** Universiti Teknologi PETRONAS, Malaysia

**Jose Ignacio Alonso** Polytechnic University of Madrid, Spain

**Reem Alsabban** University of Jeddah, Saudi Arabia

**Sultan Al-Salem** KISR, Kuwait

**Andrea Antonucci** University of Roma 3, Italy

**Srazali Aripin** International Islamic University Malaysia, Malaysia

**Eman Assi** American University of Ras Al Khaimah, UAE

**Sahar Attia** Cairo University, Egypt

**Jihad Awad** Ajman University, UAE

**Warren Axelrod** C. Warren Axelrod LLC, USA

**Mohammed Bagader** Umm Al-Qura University, Saudi Arabia

**Azizi Bahauddin** Universiti Sains Malaysia, Malaysia

**Francine Baker** Wolfson College, UK

**Marco Baldi** Marche Polytechnic University, Italy

**Michael Barber** University of Utah, USA

**Socrates Basbas** Aristotle University of Thessaloniki, Greece

**Joao Batista** University of São Paulo, Brazil

**Gianfranco Becciu** Politecnico di Milano, Italy

**Michael Beer** Leibniz Universitat Hannover, Germany

**Khadija Benis** c5Lab, Portugal

**Marco Bietresato** Free University of Bolzano, Italy

**Alma Bojorquez-Vargas** Autonomous University of San Luis Potosí, Mexico

**Daniel Bonotto** UNESP, Brazil

**Colin Booth** University of the West of England, UK

**Carlos Borrego** University of Avoiro, Portugal

**Bouzid Boudiaf** Ajman University, UAE

**Zuzana Boukalova** VODNÍ ZDROJE, a.s., Czech Republic

**Djamel Boussaa** Qatar University, Qatar

**Roman Brandtweiner** Vienna University of Economics and Business, Austria

**Roger Brewster** Bond University, Australia

**Andre Buchau** University of Stuttgart, Germany

**Raul Campos** RCQ Structural Engineering, Chile

**Richard Carranza** Carranza Consulting, USA

**Paul Carrion** Mero ESPOL Polytechnic University, Ecuador

**Joao-Manuel Carvalho** Universidade de Lisboa, Portugal

**Ana Cristina Paixao Casaca** Isel, Portugal

**Ricardo Castedo** Universidad Politécnica de Madrid, Spain

**Robert Cerny** Czech Technical University Prague, Czech Republic

**Camilo Cerro** American University of Sharjah, UAE

**Vicent Esteban Chapapria** Polytechnic University of Valencia, Spain

**Galina Chebotareva** Ural Federal University, Russia

**Hai-Bo Chen** University of Science & Technology of China, China

**Jeng-Tzong Chen** National Taiwan Ocean University, Taiwan

**Weiqiu Chen** Zhejiang University, China

**Rémy Chevrier** SNCF Innovation & Research, France

Jaroslaw Chudzicki Warsaw University of
Technology, Poland

Graham Coates Newcastle University, UK

Marcelo Enrique Conti Sapienza University
of Rome, Italy

Maria Vittoria Corazza Sapienza University
of Rome, Italy

Nelson Cordeiro CEFET-RJ, Brazil

Carlos Cuadra Akita Prefectural University,
Japan

Carmela Cucuzzella Concordia University,
Canada

Maria da Conceicao Cunha University of
Coimbra, Portugal

Luca D'Acierno University of Naples
Federico II, Italy

Boris Davydov Far Eastern State Transport
University, Russia

Miguel De Luque Pontifical Xavierian
University, Colombia

Abel Diaz Skidmore, Owings & Merrill,
USA

Jacobo Diaz University of A Coruna, Spain

Eleni Didaskalou University of Piraeus,
Greece

Petia Dineva Bulgarian Academy of
Sciences, Bulgaria

Gianluca Dini University of Pisa, Italy

Gulsen Disli Necmettin Erbakan University,
Turkey

Eduardo Divo Embry-Riddle Aeronautical
University, USA

Hrvoje Dodig University of Split, Croatia

Alexey Domnikov Ural Federal University,
Russia

Chunying Dong Beijing Insitute of
Technology, China

Mauro Dujmovic Juraj Dobrila University
of Pula, Croatia

Ney Dumont Pontifical Catholic University
of Rio de Janeiro, Brazil

Marwan El Mubarak United Arab Emirates
University, UAE

Arturo Vicente Estruch-Guitart Polytechnic
University of Valencia, Spain

Gianfranco Fancello University of Cagliari,
Italy

Alessandro Farina University of Pisa, Italy

Israel Felzenszwalb Universidade do
Estado do Rio de Janeiro, Brazil

Joana Ferreira University of Aveiro,
Portugal

Ales Filip University of Pardubice, Czech
Republic

Giulia Forestieri Universidad de La Sabana,
Colombia

Zhuojia Fu Hohai University, China

Andrew Furman Ryerson University,
Canada

Antonio Galiano Garrigos University of
Alicante, Spain

Alexander Galybin IPE RAS, Russia

Xiao-Wei Gao Dalian University of
Technology, China

Aitor Baldomir Garcia University of A
Coruna, Spain

Michael Garrison University of Texas at
Austin, USA

Gargi Ghosh Sky Group, India

Eric Gielen Polytechnic University of
Valencia, Spain

Cristina Olga Gociman Uauim University of
Architecture and Urbanism Ion Mincu
Bucharest, Romania

Luis Godinho University of Coimbra,
Portugal

Jose Antonio Souto Gonzalez University
of Santiago of Compostela, Spain

Alejandro Grindlay University of Granada,
Spain

Ove Tobias Gudmestad University of
Stavanger, Norway

Jabulani Gumbo University of Venda,
South Africa

David Hanson HansonRM, USA

Kabila Hmood Al Zaytoonah University of
Jordan, Jordan

Stanislav Hodas University of Zilina,
Slovakia

Matthieu Horgnies Holcim Innovation
Center, France

Jong-Gyu Hwang Korea Railroad Research
Institute, South Korea

Syed Ihtsham-ul-Haq Gilani University of
Technology Petronas, Malaysia

Rosaria Ippolito University of Rome "La
Sapienza", Italy

Tadaharu Ishikawa Tokyo Institute of
Technology, Japan

Malgorzata Iwanek Lublin University of
Technology, Poland

Libor Izvolt University of Zilina, Slovakia

Sharmila Jagadisan VIT Vellore, India

Yogesh Jaluria Rutgers University, USA

Gerardus Janszen Polytechnic University of Milan, Italy

Bryan Jenkins University of Adelaide, Australia

Pushpa Jha Sant Longowal Institute of Engineering & Technology, India

Zhen-Gang Ji Bureau of Ocean Energy Management, USA

Edward Kansa Convergent Solutions, USA

Andreas Karageorghis University of Cyprus, Cyprus

Alain Kassab University of Central Florida, USA

Hiroshi Kato Hokkaido University, Japan

Teruomi Katori Nihon University, Japan

Kostas Katsifarakis Aristotle Univ of Thessaloniki, Greece

John Katsikadelis National Technical University, Greece

Lina Kattan University of Jeddah, Saudi Arabia

Bashir Kazimee Washington State University, USA

Tamara Kelly Abu Dhabi University, UAE

Arzu Kocabas MSFAU, Turkey

Dariusz Kowalski Lublin University of Technology, Poland

Piotr Kowalski AGH University of Science and Technology, Poland

Mikhail Kozhevnikov Ural Federal University, Russia

Stojan Kravanja University of Maribor, Slovenia

Hrvoje Krstic Josip Juraj Strossmayer University of Osijek, Croatia

Suren Kulshreshtha University of Saskatchewan, Canada

Amaranath Sena Kumara Safetec, Norway

Shamsul Rahman Mohamed Kutty Universiti Teknologi PETRONAS, Malaysia

Lien Kwei Chien National Taiwan Ocean University, Taiwan

Jessica Lamond University of the West of England, UK

Ting-I Lee National Chiayi University, Taiwan

Vitor Leitao Universidade de Lisboa, Portugal

Daniel Lesnic University of Leeds, UK

Leevan Ling Hong Kong Baptist University, Hong Kong

Christoph Link Austrian Energy Agency, Austria

Danila Longo University of Bologna, Italy

Regina Longo Pontifical Catholic University of Campinas, Brazil

Myriam Lopes University of Aveiro, Portugal

Carlos Lopez Flanders Make VZW, Belgium

Maria Eugenia Lopez Lambas Polytechnic University of Madrid, Spain

Julia Lu Ryerson University, Canada

Aida Macerinskiene Vilnius University, Lithuania

Jussara Socorro Cury Maciel CPRM - Geological Survey of Brazil, Brazil

Isabel Madaleno University of Lisbon, Portugal

Roberto Magini Sapienza, University of Rome, Italy

Nader Mahinpey University of Calgary, Canada

Robert Mahler University of Idaho, USA

Irina Malkina-Pykh St Petersburg State Institute of Psychology and Social Work, Russia

Ulo Mander University of Tartu, Estonia

Florica Manea Politehnica University of Timisoara, Romania

George Manolis Aristotle University of Thessaloniki, Greece

Mariana Marchioni Politecnico di Milano, Italy

Liviu Marin University of Bucharest, Romania

Guido Marseglia University of Seville, Italy

Pascual Martí-montrull Technical University of Cartagena, Spain

Toshiro Matsumoto Nagoya University, Japan

Gerson Araujo de Medeiros São Paulo State University, Brazil

Sabrina Meneghello Ca' Foscari University, Italy

Paul Carrion Mero CIPAT-ESPOL, Ecuador

Ana Isabel Miranda University of Aveiro, Portugal

Hawa Mkwela Tumaini University, Tanzania

Mohamed Fekry Mohamed Effat University, Saudi Arabia

Giuseppe Musolino Mediterranea University of Reggio Calabria, Italy

Juraj Muzik University of Zilina, Slovakia
Richard Mwaipungu Sansutwa Simtali Ltd, Tanzania
Shiva Nagendra Indian Institute of Technology Madras, India
Fermin Navarrina University of A Coruña, Spain
Norwina Mohd Nawawi International Islamic University Malaysia, Malaysia
Derek Northwood University of Windsor, Canada
David Novelo-Casanova National Autonomous University of Mexico, Mexico
Andrzej Nowak Silesian University of Technology, Poland
Freeman Ntuli Botswana International University of Science and Technology, Botswana
Miguel Juan Nunez-Sanchez European Maritime Safety Agency, Portugal
Suk Mun Oh Korea Railroad Research Institute, South Korea
Yasuo Ohe Tokyo University of Agriculture, Japan
Roger Olsen CDM Smith, USA
Antonio Romero Ordonez University of Seville, Spain
Francisco Ortega Riejos Universidad de Sevilla, Spain
Ozlem Ozcevik Istanbul Technical University, Turkey
Leandro Palermo Jr University of Campinas, Brazil
Deborah Panepinto Turin Polytechnic, Italy
Marilena Papageorgiou Aristotle University of Thessaloniki, Greece
Jose Paris University of A Coruna, Spain
Bum Hwan Park Korea National University of Transportation, South Korea
Bekir Parlak Bursa Uludag University, Turkey
Rene Parra Universidad San Francisco de Quito, Ecuador
Marko Peric University of Rijeka, Croatia
Roberto Perruzza CERN, Switzerland
Cristiana Piccioni Sapienza University of Rome, Italy
Max Platzer AeroHydro Research & Technology Associates, USA
Lorenz Poggendorf Toyo University, Japan
Dragan Poljak University of Split, Croatia

Antonella Pontrandolfi Council for Agricultural Research & Economics, Italy
Aniela Pop Polytechnic University of Timisoara, Romania
Serguei Potapov French Electricity (EDF), France
Maria Pregnolato University of Bristol, UK
Dimitris Prokopiou University of Piraeus, Greece
Yuri Pykh Russian Academy of Sciences, Russia
Sue Raftery OnPoint Learning, USA
Marco Ragazzi University of Trento, Italy
Marco Ravina Turin Polytechnic, Italy
Jure Ravnik University of Maribor, Slovenia
Joseph Rencis California State Polytechnic University, USA
Genserik Reniers University of Antwerp, Belgium
Admilson Irio Ribeiro São Paulo State University, Brazil
Jorge Ribeiro University of Lisbon, Portugal
Angelo Riccio University of Naples "Parthenope", Italy
Corrado Rindone Mediterranea University of Reggio Calabria, Italy
German Rodriguez Universidad de Las Palmas de Gran Canaria, Spain
Rosa Rojas-Caldelas Autonomous University of Baja California, Mexico
Jafar Rouhi University of Campania "L. Vanvitelli", Italy
Irina Rukavishnikova Ural Federal University, Russia
Francesco Russo University of Reggio Calabria, Italy
Shahrul Said Universiti Teknologi MARA, Malaysia
Hidetoshi Sakamoto Doshisha University, Japan
Seddik Sakji INFRANEO, France
Artem Salamatov Chelyabinsk State University, Russia
Daniel Santos-Reyes ICHI Research & Engineering, Mexico
Bozidar Sarler University of Ljubljana, Slovenia
Martin Schanz Graz University of Technology, Austria
Evelia Schettini University of Bari, Italy

Marco Schiavon University Of Trento, Italy

Michal Sejnoha Czech Technical University, Czech Republic

Marichela Sepe University of Naples Federico II, Italy

Angela Baeza Serrano Global Omnium Medioambiente, S.L., Spain

Leticia Serrano-Estrada University of Alicante, Spain

Wael Shaheen Palestine Polytechnic University, Palestine

Viktor Silbermann Fichtner GmbH & Co KG, Germany

Luis Simoes University of Coimbra, Portugal

Nuno Simoes University of Coimbra, Portugal

Sradhanjali Singh CSIR-NEERI, Delhi, India

Rolf Sjoblom Luleå University of Technology, Sweden

Leopold Skerget Slovenian Academy of Engineering, Slovenia

Vladimir Sladek Slovak Academy of Sciences, Slovakia

Alexander Slobodov ITMO University, Russia

Lauren Stewart Georgia Institute of Technology, USA

Elena Strelnikova National Academy of Sciences of Ukraine, Ukraine

Michel Olivier Sturtzer French-German Research Institute of Saint-Louis, France

Miroslav Sykora Czech Technical University of Prague, Czech Republic

Antonio Tadeu University of Coimbra, Portugal

Paulo Roberto Armanini Tagliani Federal University of Rio Grande, Brazil

Hitoshi Takagi Tokushima University, Japan

Kenichi Takemura Kanagawa University, Japan

Manal Yehia Tawfik El Shourok Academy, Egypt

Filipe Teixeira-Dias University of Edinburgh, UK

Roberta Teta University of Naples Federico II, Italy

Norio Tomii Nihon University, Japan

Juan Carlos Pomares Torres University of Alicante, Spain

Vincenzo Torretta Insubria University, Italy

Sophie Trelat IRSN, France

Carlo Trozzi Teche Consulting srl, Italy

Sirma Turgut Yildiz Technical University, Turkey

Charles Tushabomwe-Kazooba Mbarara University of Science & Technology, Uganda

Elen Twrdy University of Ljubljana, Slovenia

Maria Valles-Panells Polytechnic University of Valencia, Spain

Thierry Vanelslander University of Antwerp, Belgium

Baxter Vieux Vieux & Associates, Inc., USA

Antonino Vitetta Mediterranea University of Reggio Calabria, Italy

Jaap Vleugel Delft University of Technology, Netherlands

Giuliano Vox University of Bari, Italy

Adam Weintritt Gdynia Maritime University, Poland

Alvyn Williams Soft Loud House Architects, Australia

Ben Williams University of the West of England, UK

Shahla Wunderlich Montclair State University, USA

Wolf Yeigh University of Washington, USA

Victor Yepes Universitat Politecnica de Valencia, Spain

Montserrat Zamorano University of Granada, Spain

Giuseppe Zappala National Research Council, Italy

Chuanzeng Zhang University of Siegen, Germany

Dichuan Zhang Nazarbayev University, Kazakhstan

# Sustainable Tourism X

**Editors**

**Stavros Syngellakis**
*Wessex Institute, UK*
*Member of WIT Board of Directors*

**Pablo Diaz Rodriguez**
*University of La Laguna, Spain*

**WIT**PRESS  Southampton, Boston

**Editors:**

**Stavros Syngellakis**
*Wessex Institute, UK*
*Member of WIT Board of Directors*

**Pablo Diaz Rodriguez**
*University of La Laguna, Spain*

Published by

**WIT Press**
Ashurst Lodge, Ashurst, Southampton, SO40 7AA, UK
Tel: 44 (0) 238 029 3223; Fax: 44 (0) 238 029 2853
E-Mail: witpress@witpress.com
http://www.witpress.com

For USA, Canada and Mexico

**Computational Mechanics International Inc**
25 Bridge Street, Billerica, MA 01821, USA
Tel: 978 667 5841; Fax: 978 667 7582
E-Mail: infousa@witpress.com
http://www.witpress.com

British Library Cataloguing-in-Publication Data

A Catalogue record for this book is available
from the British Library

ISBN: 978-1-78466-461-9
eISBN: 978-1-78466-462-6
ISSN: 1746-448X (print)
ISSN: 1743-3541 (on-line)

*The texts of the papers in this volume were set individually by the authors or under their supervision. Only minor corrections to the text may have been carried out by the publisher.*

No responsibility is assumed by the Publisher, the Editors and Authors for any injury and/or damage to persons or property as a matter of products liability, negligence or otherwise, or from any use or operation of any methods, products, instructions or ideas contained in the material herein. The Publisher does not necessarily endorse the ideas held, or views expressed by the Editors or Authors of the material contained in its publications.

© WIT Press 2022

Open Access: All of the papers published in this journal are freely available, without charge, for users to read, download, copy, distribute, print, search, link to the full text, or use for any other lawful purpose, without asking prior permission from the publisher or the author as long as the author/copyright holder is attributed. This is in accordance with the BOAI definition of open access.

Creative Commons content: The CC BY 4.0 licence allows users to copy, distribute and transmit an article, and adapt the article as long as the author is attributed. The CC BY licence permits commercial and non-commercial reuse.

# Preface

This book contains a selection of papers among those presented at the 10th International Conference on Sustainable Tourism, organized by the Wessex Institute of Technology, UK and University of La Laguna, Spain. The meeting was sponsored by the WIT Transactions on Ecology and the Environment, the International Journal of Environmental Impacts and the International Journal of Energy Production and Management.

The contributions to this book address a wide range of topics related to climate change, ecotourism, marine and coastal tourism, rural tourism, cultural tourism, tourism and technology as well as assessments of tourism sustainability.

The effects of climate change can be much more visible in travel regions and need to be continuously monitored and evaluated so that necessary measures can be taken. Such measures can be identified at a regional level through communication with stakeholders in the tourism industry and tourism destinations. Additionally, data on climate change made available via a climate information system serve to raise awareness on this issue among stakeholders.

Sustainable design strategies of hotels and resorts located on fragile ecosystems can be adopted with the aim of reversing climate change and addressing the encroachments on nature and mishandling of waste. Such a strategy would involve full integration of a resort into the natural landscape, natural cooling and airing of the spaces, natural and recyclable construction materials, renewable energy sources, use of local vegetation, water recirculation, waste processing, and garbage recycling.

Diving in kelp is a recreational activity that has the potential to become a major part of marine tourism in certain countries. Management of kelp diving as marine tourism activity includes better patrol of kelp forests, codes of conduct for divers, education, and marketing of sustainable kelp diving and other non-diving activities. This can be guided by gathering data on the demographic, psychographic, behavioural and specialisation profiles of kelp divers.

The incorporation of traditional fishing to the list of tourist experiences can improve the living conditions of fishing populations and reduce the pressure on fishery resources. This would require strategies for the participation of the different groups of agents involved in the activity. More specifically, training and awareness strategies for the incorporation of fishermen as tourism service providers.

The features of potential demand targets, required rural attractions, and essential facilities need to be identified for rural tourism planning and development. The natural environment, rural heritage, and local food are three common major attractions for two types of rural tourism, namely, micro-tourism, that is, tourism around tourists' neighbouring areas, and workcation, that is, the combination of remote work and vacation.

New forms of spiritual tourism can lead to an expanded understanding of sustainability. By adding spiritual tourism to the concept of sustainable tourism, it would be possible to integrate the spiritual dimension of human existence into an even more fulfilling travel experience. Balance in the outer world would be achieved through balance in our inner world.

Establishment of contacts with people from different countries and cultures often is a very important motivation for travelling. Social interaction and effective intercultural communication in tourism encounters may result in mutual appreciation, understanding, respect, tolerance and the overall improvement of the social interactions between individuals, but such encounters may also represent a potential minefield due to cultural differences in communication and rules of social behaviour.

Festivals can be an instrument for tourism development, city image improvement and boosting regional economies. Taking into account the views of the leading players of tourism, local residents and business owners can enrich the existing knowledge on promoting sustainable events and sustainable approaches to tourism development. The results may be useful not only to local government entities involved in the tourism strategic planning but also to stakeholders engaged in creating sustainable competitive advantage in the tourism industry.

Due to the widespread digitization, the tourism industry is overwhelmed by a huge amount of data that needs to be processed and analysed. The adoption of Artificial Intelligence, Big Data Analytics and Smart Tourism may enable tourist companies to increase their business performance, achieve economic results and potentially attain a sustainable competitive advantage.

For the promotion of sustainable tourism, it is necessary to develop a tool capable of assessing the impacts associated with the tourist sector and to identify which actions are currently being taken in order to achieve the desired level of sustainability. Life Cycle Assessment is a highly effective tool capable of assessing direct and indirect carbon emissions as well as the socioeconomic and environmental impacts generated in the tourism sector. Eco-labelling and digitalisation of the tourism experiences are also valuable tools to minimize negative environmental impacts, to promote mechanisms to access green markets and to frame successful synergies.

Various environmentally sustainable practices (ESP) are used by hoteliers which however have been changing according to the guests' expectations; it is therefore important, to identify the gap between guests' expectations and satisfaction of ESP. There are factors for which this gap is significant and solutions need to be found so that ESP meet the satisfaction level of guests.

Sustainable models in hospitality may be promoted by the spread of a procurement strategies based on local producers. Such exemplary sustainable practices may be applied and communicated by high-end restaurants.

This volume is part of the WIT Transactions on Ecology and the Environment. The digital version of the papers, as well as those presented in the previous conferences are archived in Open Access format in the eLibrary of the Wessex Institute (https://www.witpress.com/elibrary) where they are freely available to the international community.

The editors are grateful to all authors for the quality of their contributions as well as to the members of the International Scientific Advisory Committee and other colleagues who helped to review the papers and hence ensure the quality of this volume.

The Editors, 2022

# Contents

## Section 4: Tourism and sustainability

# SECTION 1
# TOURISM AND
# CLIMATE CHANGE

# IMPACTS OF CLIMATE CHANGE ON TOURISM AREAS IN GERMANY: AN OVERVIEW OF RECOMMENDED ADAPTATION MEASURES AND THEIR COMMUNICATION

ULRIKE WACHOTSCH*
International Sustainability Strategies, Policy and Knowledge Transfer, German Environment Agency, Germany

## ABSTRACT

The climate in Germany will change due to human-induced climate change. Therefore, the German Environment Agency is continuously monitoring and evaluating these changes on behalf of the Federal Government in order to advise it on the necessary measures. Climate change will also affect the tourism industry and travel regions in Germany. Hence, temporary or permanent restrictions on business activities may occur, endangering jobs and incomes. The changes are much more visible on the level of the travel regions than on the national level. In a research project, measures for Germany that can be taken on the regional level to counter the consequences were identified through communication with stakeholders in the tourism industry and tourism destinations. Additionally, data on climate change are made available via a climate information system. This serves to raise awareness of climate change among stakeholders.

*Keywords: sustainable tourism, climate adaptation, travel region, policy advice, communication, adaptation measure.*

## 1 INTRODUCTION

It is known that the findings on possible climate change and its causes date back to around 1965. It was already recognised by scientists at that time that carbon dioxide emissions were responsible for climate changes. Thereafter, the first climate models were developed in the 1970s and experts pointed out the need for changes in economic and consumer behaviour in the media or advisory bodies. In 1958, the German poet Hans Magnus Enzensberger pointed out in an essay that tourism, and especially mass tourism, has a great destructive potential [1]. Since then, tourism has grown many times over and with it the pressure on the environment and nature. If you read scientific papers or political concepts on sustainable tourism, many different topics are addressed, for example energy use, waste and water management, fair working conditions, inclusion or economic success. Daniel Scott argues the following in 2010: "Tourism is currently considered among the economic sectors least prepared for the risks and opportunities posed by climate change and is only now developing the capacity to advance knowledge necessary to inform businesses, communities and government about the issues and potential ways forward" [2]. The concern with climate change and adaptation to its consequences is continuing to this day [3].

The economic contribution of the German tourism industry amounts to 3.9% of the gross domestic product and secures a total of 2.9 million jobs [4]. Both the economic and recreational factors of tourism are influenced by the effects of global climate change. The tourism sector itself also contributes to climate change: Globally, the World Tourism Organization (UNWTO) and the United Nations Environment Programme estimated tourism's contribution to climate change at between 4.6% and 7.8% [5]. Similar values are also included in the report of the Intergovernmental Panel on Climate Change [6]. In addition,

* *ORCID: https://orcid.org/ 0000-0002-8042-1360*

WIT Transactions on Ecology and the Environment, Vol 256, © 2022 WIT Press
www.witpress.com, ISSN 1743-3541 (on-line)
doi:10.2495/ST220011

the current findings of the IPCC show that with the climate-damaging emissions we are already emitting, we can expect global warming of 1.5 or 2 degrees, and even more without a change in our consumption and economic practices. This article is not about analysing who would have to contribute what amount to reduce tourism's contribution to climate change. There are already a number of agreements and recommendations on this. However, there is a lack of comprehensive and ambitious implementation of measures to achieve climate protection goals in the tourism industry [7]. Therefore, the purpose of this article is to address the problem that climate changes will occur in all regions of the world. These will have an influence on the economy and thus also on the tourism industry. In the following, the article will look at the situation in Germany.

## 2  THE SITUATION IN GERMANY

The vulnerability analysis of Germany to climate change published in 2015 deals with the tourism industry as an area for action. The study identifies three main impacts of climate change on the tourism industry through an impact chain analysis. These are (1) business interruptions; (2) seasonal and regional shifts in demand; and (3) climate-related demands on tourism infrastructures. Overall, the impact on the tourism industry is classified as low in relation to other fields of action. In the event of a strong change in the climate, a change would most likely occur in the area of tourism infrastructures, according to the study's findings. At the same time, the tourism industry is considered to have a high capacity for adaptation. However, the affectedness and adaptation needs of individual tourism segments differ greatly. The climate signals of river floods, storm surges, flash floods, temperature, humidity, heat, precipitation, snowfall and changes in radiation were identified by the study as particularly relevant for the tourism industry [8]. Questions of changing conditions for travel are addressed in the context of ski tourism in the Alps and low mountain ranges and partly in connection with coastal regions [9]–[11].

There is a scientific consensus that tourism is influenced by the climate, as many tourism activities take place in nature or in front of the landscape. At the same time, the exact influence of climate change is hardly quantifiable for Germany due to the lack of data: which is mostly available on a monthly basis. Quantification would require trip or turnover data on a daily basis. A case study for an Alpine destination can already establish a connection by analysing a combination of weather data and overnight stay data [12].

In 2008, the Federal Government launched the German Strategy for Adaptation to Climate Change [13]. This strategy identifies the need for change and adaptation in tourism, particularly in coastal areas and mountain regions. The adaptation strategy is accompanied by a reporting and monitoring process [14]. The monitoring system is regularly evaluated and improved [15]. For this purpose, indicators were developed that show both the influence of the changes caused by climate change on the tourism industry ("impact indicators") and its reactions ("response indicators").

The Climate Impact and Risk Assessment 2021 for Germany specifies the risks for the tourism field of action in the case that no adaptation takes place (Fig. 1).

The report points out that the impact of climate change varies from region to region. It can be seen that most of the tourism sectors under consideration are currently only slightly affected. It is expected that this will increase as climate change progresses. Due to the multicausality of regional differences in geography, climate and tourism offerings, it is hard to describe impacts and adaptation capacities more precisely on the national level.

"Official statistics distinguish 143 travel areas in Germany, which usually correspond to a district in terms of their spatial layout" [17]. The German Environment Agency has

| Climate impact | | Present | Middle of the century | | End of the century | | Adaptation period |
|---|---|---|---|---|---|---|---|
| | | | optimistic | pessimistic | optimistic | pessimistic | |
| Restriction of tourism options: effects of a lack of guaranteed snow on winter tourism | Climate risk | low | medium | medium | high | high | < 10 years |
| | Certainty | | medium | | medium | | |
| Restriction of tourism options: effects of heat on health-based tour-ism | Climate risk | low | low | medium | medium | medium | < 10 years |
| | Certainty | | low | | low | | |
| Damage to tourist infra-structure and business interruptions | Climate risk | low | medium | medium | medium | high | < 10 years |
| | Certainty | | medium | | low | | |
| Shift in demand | Climate risk | low | low | medium | low | medium | < 10 years |
| | Certainty | | low | | low | | |
| Economic opportunities and risks for tourism | Climate risk | medium | medium | medium | high | high | < 10 years |
| | Certainty | | medium | | low | | |

Figure 1:  Climate risks without adaptation in the tourism action field [16].

launched an investigation with the title Impacts of climate change on tourism in 2017. The study aimed to show how tourism regions are and will be affected by changes in climate. By providing information, data and suggestions for measures, destination management organizations should be enabled to implement adaptation measures themselves. The research was implemented jointly by several institutes and was completed in 2021 [17].

## 3 ANALYSIS ON THE LEVEL OF TOURISM DESTINATIONS

In relation to the difficulties of accurately describing the consequences of climate change on the national level, it is useful to look at the regional level. The following results are available in the research project "Impacts of climate change on tourism in the German alpine and low mountain ranges and coastal regions as well as on bathing tourism and river-related forms of tourism (e.g. cycle and water tourism)" [17]. The key question is which changes in the climate will bring about the need for adaptation for the tourism industry. Both extreme weather events and changes in the general conditions that occur gradually and then remain are relevant to consider. Extreme events in Germany are storms, floods, heavy rain, forest fires, low water and drought. The general conditions are influenced by changes in temperature or the amount of precipitation. Depending on the region, this leads to further consequences, especially with regard to nature or people's health and well-being. "Since weather records began in 1880, the air temperature in Germany has increased by 1.4°C, with a particularly sharp rise since the 1960s. Extreme values are also increasing, with heat days occurring twice as often today as in the climate reference period 1961–1990. This particularly affects the regions of Rheinhessen and Spreewald. Heavy rainfall events are occurring with increasing frequency, especially in northern and central Germany. Ice days, which occur most frequently in the high and low mountain regions, decrease significantly. On average across Germany, the number of ice days decreases by about 1 day per 10 years" [17].

The research team first identified the geographical spread of the 143 tourism areas into which Germany is divided. For these regions, statistics are available on the number of guest arrivals and length of stay on a monthly basis. In the next step, recognised climate models were used to calculate the expected change in the number of visitors. The "observed climatic changes of the last decades (1961–2019) provide information on which developments we should expect in the near future and allow a classification of regional climate projections for medium time horizons beyond 2050. In order to gain insights into the future climatic development, 4 regional climate simulations were evaluated for the emission scenario RCP8.5 ('business-as-usual') (see chapter 3.2.2). A climate information system […] was developed for the illustrative presentation of various climatic parameters, which describes past and possible future climatic development using interactive maps, tables, and time series" [17] (Fig. 2). The climate information system makes it possible to call up and display climate changes either for a selected travel region or in the overview for Germany with all regions.

Each of the climate changes will require tourism offers to be adapted. Regionally, it may also be that certain activities are temporarily or permanently not possible. In relation to increasing drought, this is associated with an increased risk of forest fires, impairment of biodiversity and water shortages. Water is used for various tourism activities, which can then be restricted. These include the production of artificial snow, but also canoeing and swimming. But also the use of streams for energy generation, e.g. in small hydroelectric power plants in mountain regions, may be limited. Impairment of the forest in turn affects activities that are carried out in the forest, such as hiking, geocaching or cycling.

## 4 ADAPTATION MEASURES

The German Environment Agency presents a number of adaptation measures on its website. However, there was a lack of specific recommendations for the tourism industry and related tourism activities. After the analysis of the changes on the regional level, the commissioned research team was able to start working on adaptation measures. "In order to identify possible adaptation measures, a scheme was developed to analyse changes due to climate change and impacts on tourism elements and forms. Based on a literature review and four expert workshops, 24 measures were selected for closer examination" [17].

The measures could be divided into the following six categories: (1) Leisure activities independent of weather conditions; (2) Technical measures in the event of changes in the general conditions; (3) Crisis management; (4) Crisis prevention; (5) Product and marketing adaptation; and (6) Guidance of visitor flow [17]. The measures described must be selected on the basis of the changes to be expected regionally and relate to the existing offers of the travel area.

Table 1 shows all climate change adaptation measures that were found for German travel areas. All measures are presented and assessed in detail to help local actors implement and integrate them into their own work. For each measure, a description is given of the climate impacts it can improve. A measure can address several changes, as it is complex. Furthermore, for each measure there is an indication of the steps that a person responsible for implementation or the team must undertake in order to implement it. In addition, obstacles and their solutions are already pointed out as far as foreseeable. Costs often play an important role in implementation, so what can be expected is discussed. This is described qualitatively, as it is not possible to make a quantitative statement, because too many factors are unknown, such as the size of the region, previous work that can be drawn upon, etc. Furthermore, almost none of these measures have been put into practice so far, so that no empirical values can be used. Moreover, there are estimates of the ecological and socio-economic impact of each

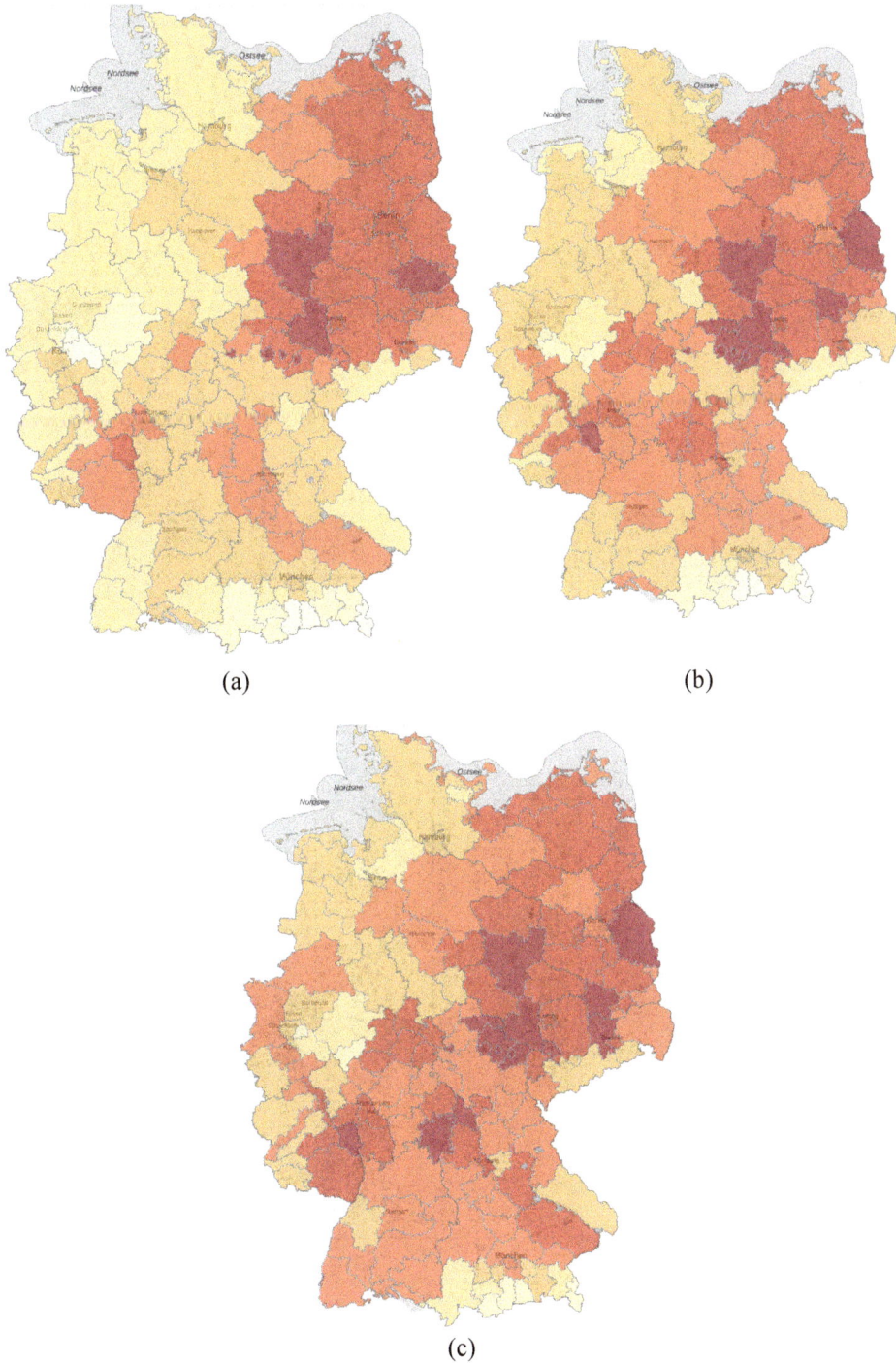

(a)

(b)

(c)

Figure 2:    The number of dry days per year in comparison to three time periods. (a) 1990–2019; (b) 2031–2060; and (c) 2071–2100 [19].

Table 1: Overview of the selected measures [17].

| Category of adaptation measure | Specific adaptation measure | | | |
|---|---|---|---|---|
| Leisure activities independent of weather conditions | Shading of paths against increasing heat | | Footbridges that adapt to water levels | |
| Technical measures in the event of changes in the general conditions | Creation of water areas | Measures to save water in tourist facilities | Water-saving outdoor facilities | Creating more recreational facilities that are not dependent on weather conditions |
| Crisis management | Establish and refine crisis management | | Create evacuation and communication concepts | |
| Crisis prevention | Consistently and systematically monitor endangered areas that are considered tourist areas | Carry out risk analyses and natural hazard scenarios for tourism, continuously update risk mapping | Promote learning cooperation for mutual support in hazard prevention | Impose requirements on tourism infrastructure in order to be better equipped against hazard scenarios |
| | Raise awareness of climate change among tourism service providers and associations. | Educate the population about weather risks and natural hazards – Inform guests openly | Train staff to save water and energy | Train staff on how to behave in crisis situations |
| Product and marketing adaptation | Modification/tightening of cancellation conditions | Risk minimisation for (large) events | Product adaptation | |
| Guidance of visitor flow | Guidance through information and targeted offers | | Guidance through bans and restrictions | |

measure. Not all measures could be recommended from an ecological point of view, e.g. the use of artificial snow in skiing regions that can expect less and less snowfall [17].

## 5  ACCOMPANYING COMMUNICATION ON RESEARCH

The study "Impacts of climate change on tourism in the German alpine and low mountain ranges and coastal regions as well as on bathing tourism and river-related forms of tourism (e.g. cycle and water tourism)" [17] was commissioned with the aim of providing policy advice. It was important from the start that the results should be used for practical application. Another aspect to consider is that the awareness of the need for adaptation of the tourism industry to climate change was not very high at the beginning of the project. Furthermore, measures for future implementation were developed without being able to look back on an extensive evaluation of similar measures in the past, as climate change is occurring for the first time. These facts required extensive communication with local actors in the travel areas. The aim of the communication was firstly to include existing knowledge, secondly to identify needs, thirdly to prepare research results in a target group-oriented and practical manner, fourthly to raise awareness of the need for adaptation measures and fifthly to encourage implementation.

In the different phases of the research project, communication with local stakeholders of the tourism industry varied (Fig. 3). After the first analysis of available data, a workshop was offered. The aim was to narrow down which climate changes are particularly relevant for tourism and which are not. Furthermore, what support is needed for adaptation on the local level was identified and discussed. It became clear that a guide summarising the information would be helpful. It should present the impacts of climate change, provide guidance on how to start and implement an adaptation process, how to finance it and what measures can be taken. In addition, the preparation of data on the level of the tourism region was considered helpful by the stakeholders.

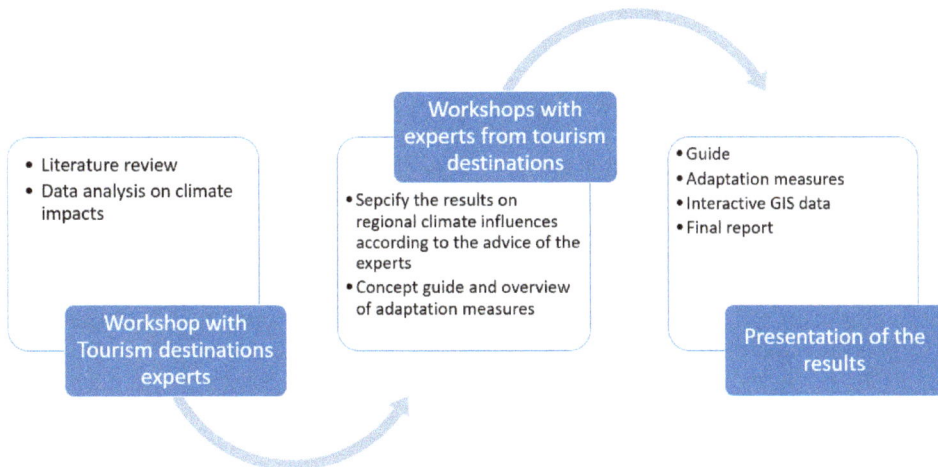

Figure 3:  Process of the contracted research project and the accompanying communication on the topic with experts from the tourism regions.

In the next step, a guideline [18] was developed that responds to the indications. This results in the contents: Presentation of the necessity of adaptation to climate change, description of the expected changes. It also describes how an adaptation process can be started. It then goes on to describe which adaptation measures are possible and recommended, and how financial resources can be found. After another round of region-specific workshops with stakeholders, the guide was finalised, the data was entered into a GIS climate information system [19] and the adaptation measures [20] were presented on the website of the German Environment Agency. While the guide is available free of charge in pdf and printed form, the pdf format can only be accessed online. The reason for this is that they can be updated quickly if new findings emerge. All results were presented to the interested expert public during the first lockdown in an online conference with almost 300 participants [21]. In addition, two brochures were produced on the topics of how to use the climate information system [22] and climate protection versus climate adaptation – where are the differences? [23]. Finally, the scientific report was published and a press release was issued.

## 6 OUTLOOK

Were the objectives of the project achieved? Yes, many of the objectives have been achieved, it is now clear that climate changes will occur in tourism regions that will require adaptation measures in the future. It is also clear what these changes will be in the individual tourism areas. Through the accompanying communication, many of the 143 travel regions have been able to achieve a raised awareness. The information materials provided have been adapted to their needs. One federal state have launched his own adaptation campaign [24] for their regions. The Excellence Initiative Sustainable Destinations in Germany has included the topic in its knowledge portal. Furthermore, the results have been noted in the scientific community.

## REFERENCES

[1]     Enzensberger, H.-M., Eine theorie des tourismus, *Merkur*, **126**, pp. 701–720, 1958.
[2]     Scott, D., Why sustainable tourism must address climate change. *Journal of Sustainable Tourism*, **19**(1), pp. 17–34, 2011.
[3]     Paquin, D., de Elía, R., Bleau, S., Charron, I., Logan, T. & Biner, S., A multiple timescales approach to assess urgency in adaptation to climate change with an application to the tourism industry. *Environmental Science and Policy*, **63**, pp. 143–150, 2016.
[4]     Rösner, J.-M., Studie Wirtschaftsfaktor Tourismus in Deutschland Kennzahlen einer umsatzstarken Querschnittsbranche. Bundesministerium für Wirtschaft und Energie, 2017. https://www.bmwk.de/Redaktion/DE/Publikationen/Tourismus/wirtschaftsfak tor-tourismus-in-deutschland-lang.html. Accessed on: 29 Apr. 2022.
[5]     Simpson, M.C., Gössling, S., Scott, D., Hall, C.M. & Gladin, E., Climate change adaptation and mitigation in the tourism sector: Frameworks, tools and practices. UNEP, University of Oxford, UNWTO, WMO, 2008. https://wedocs.unep.org/ bitstream/handle/20.500.11822/9681/Climate_Change_adaptation_mitigation.pdf?se quence=3&amp%3BisAllowed=. Accessed on: 29 Apr. 2022.
[6]     Groupe d'experts intergouvernemental sur l'évolution du climat (ed.), Climate change 2014: Mitigation of climate change. Working Group III contribution to the fifth assessment report of the Intergovernmental Panel on Climate Change. Cambridge University Press: New York, 2014.
[7]     Scott, D. & Gössling, S., From Djerba to Glasgow: Have declarations on tourism and climate change brought us any closer to meaningful climate action? *Journal of Sustainable Tourism*, **30**(1), pp. 199–222, 2022.

[8]   Dessau-Roßlau: Umweltbundesamt, Vulnerabilität Deutschlands gegenüber dem Klimawandel, 2015. http://www.umweltbundesamt.de/publikationen/vulnerabilitaet-deutschlands-gegenueber-dem.

[9]   Haanpää, S., Juhola, S. & Landauer, M., Adapting to climate change: Perceptions of vulnerability of down-hill ski area operators in southern and middle Finland. *Curr. Issues Tour.*, **18**(10), pp. 966–978, 2015.

[10]  Bonzanigo, L., Giupponi, C. & Balbi, S., Sustainable tourism planning and climate change adaptation in the Alps: A case study of winter tourism in mountain communities in the Dolomites. *J. Sustainable Tour.*, **24**(4), pp. 637–652, 2016.

[11]  Salim, E., Ravanel, L., Deline, P. & Gauchon, C., A review of melting ice adaptation strategies in the glacier tourism context. *Scandinavian Journal of Hospitality and Tourism*, **21**(2), pp. 229–246, 2021.

[12]  Bausch, T., Gartner, W.C. & Humpe, A., How weather conditions affect guest arrivals and duration of stay: An alpine destination case. *Journal of Tourism Research*, **23**(6), pp. 1006–1026, 2021

[13]  Deutsche Anpassungsstrategie an den Klimawandel, Berlin, 2008. (vom Bundeskabinett am 17. Dezember 2008 beschlossen.)

[14]  Schönthaler, K., von Andrian-Werburg, S. & van Rüth, P. (eds), Monitoringbericht 2015 zur Deutschen Anpassungsstrategie an den Klimawandel. Bericht der Interministeriellen Arbeitsgruppe Anpassungsstrategie der Bundesregierung, Dessau-Roßlau, 2015.

[15]  van Rüth, P., Schönthaler, K. & von Andrian-Werburg, S., Monitoringbericht 2019 zur Deutschen Anpassungsstrategie an den Klimawandel. Bericht der Interministeriellen Arbeitsgruppe Anpassungsstrategie der Bundesregierung. Dessau-Roßlau: Umweltbundesamt, 2019. https://www.umweltbundesamt.de/publikationen/monitoringbericht-2019.

[16]  Kahlenborn, W., Porst, L. & Voß, M., Climate Impact and Risk Assessment 2021 for Germany. Dessau-Roßlau: Umweltbundesamt, 2021. http://www.umweltbundesamt.de/publikationen/KWRA-English-Summary.

[17]  Dworak, T., Lotter, F., Hoffmann, P., Bausch, T. & Günther, W., Folgen des Klimawandels für den Tourismus in den deutschen Alpen- und Mittelgebirgsregionen und Küstenregionen sowie auf den Badetourismus und flussbegleitende Tourismusformen. Dessau-Roßlau: Umweltbundesamt, 2021. http://www.umweltbundesamt.de/publikationen/folgen-des-klimawandels-fuer-den-tourismus-in-den.

[18]  Dworak, T., Schmölzer, A. & Günther, W., Anpassung an den Klimawandel: die Zukunft im Tourismus gestalten. Dessau-Roßlau: Umweltbundesamt, 2020. http://www.umweltbundesamt.de/publikationen/anpassung-an-den-klimawandel-die-zukunft-im.

[19]  Klimawandel und Tourismus. https://gis.uba.de/maps/resources/apps/tourismus/index.html?lang=de. Accessed on: 2 May 2022.

[20]  Übersicht Anpassungsmaßnahmen im Tourismus.

[21]  Klimaszenarien für die Tourismusregionen in Deutschland. https://youtu.be/TI3WORctSZQ. Accessed on 2 May 2022.

[22]  Weisz, H., Peters, V. & Pichler, P.P., Anwendungen von Konzepten, Werkzeugen und Methoden der integrierten Risikobewertung – Entscheidungshilfen für Anpassung an den Klimawandel. Dessau-Roßlau: Umweltbundesamt, 2016. http://www.umweltbundesamt.de/publikationen/anwendung-von-konzepten-werkzeugen-methoden-der.

[23]  Lotter, F. & Dworak, T., Klimaschutz und Klimawandelanpassung. DessauRoßlau: Umweltbundesamt, 2021. http://www.umweltbundesamt.de/publikationen/faltblatt-klimaschutz-und-klimawandelanpassung.

[24]  TMN und Wirtschaftsministerium machen Niedersachsens Tourismus fit für den Klimawandel. Tourismusnetzwerk Niedersachsen.
https://nds.tourismusnetzwerk.info/inhalte/qualitaetsmanagement/klimawandel-und-tourismus/tmn-und-wirtschaftsministerium-machen-niedersachsens-tourismus-fit-fuer-den-klimawandel/. Accessed on: 5 May 2022.

# SUSTAINABLE HOTEL DESIGN STRATEGIES: TOURISM AS A TOOL FOR CIRCULAR BIOECONOMY IN FRAGILE ECOSYSTEMS

ALEXANDROS KITRINIARIS
Kitriniaris Associates Architecture Firm (KAAF), Greece
School of Architecture, National Technical University of Athens (NTUA), Greece
Institute for Advanced Architecture of Catalonia (IAAC), Spain

## ABSTRACT

This paper focuses on the exploration of sustainable design strategies of hotels and resorts located on fragile ecosystems with the aim of reversing climate change and addressing the encroachments on nature and mishandling of waste that together lead to increased risk of repeated pandemics, thus making our planet more resilient to both. The tourism industry is responsible for 5% of greenhouse gas ($CO_2$) emissions and is expected to increase by 130% by 2050. The World Tourism Organization and the United Nations Development Programme are committed to inspiring and facilitating collaboration with stakeholders to advance the contribution of tourism to the sustainable development goals. To this end, the paper explores the sustainable design strategies of the 6-star sustainable hotel and resort, Caves & Waves, as a case study. The resort is fully integrated into the natural landscape, it takes advantage of the orientation and sea view with integrity throughout all the common and private spaces, 10% of the total capacity facilitates disabled access, it favours natural cooling and airing of the spaces, it is constructed from natural and recyclable construction materials (local stone, wood), it uses renewable energy sources (solar rooves), it reinforces local vegetation, it handles water recirculation, waste processing, and garbage recycling correctly, and as a result, it is integrated into the international network of sustainable constructions by minimizing its ecological footprint, while simultaneously increasing its positive effect on the environment. Through the analysis of the above-mentioned case study, the paper highlights the key sustainable tourism design strategies regarding the 3-zero triple bottom line concept committed to zero kilometres, zero carbon dioxide and zero waste with the aim of handling the tensions between sustainable hotel design in fragile ecosystems and commitments to business development and continuing economic growth.

*Keywords: sustainable tourism, sustainable development, climate change, circular economy, ecological, hotel design strategies, fragile ecosystems.*

## 1 INTRODUCTION

In the last 50 years, the biosphere, upon which humanity depends, has been altered to an unparalleled degree. The current economic model relying on fossil resources and addicted to "growth at all costs" is putting at risk not only life on our planet, but also the world's economy. The need to react to the unprecedented COVID-19 crisis is a unique opportunity to transition towards a sustainable wellbeing economy centred around people and nature. After all, deforestation, biodiversity loss and landscape fragmentation have been identified as key processes enabling direct transmission of infectious diseases. Likewise, a changing climate has profound implications for human health. Putting forward a new economic model requires transformative policies, purposeful innovation, access to finance, risk-taking capacity as well as new and sustainable business models and markets [1].

## 2 CIRCULAR BIOECONOMY IN FRAGILE ECOSYSTEMS

A circular bioeconomy offers a conceptual framework for using renewable natural capital to holistically transform and manage our land, food, health, and industrial systems with the goal of achieving sustainable wellbeing in harmony with nature. There is a need to mainstream

WIT Transactions on Ecology and the Environment, Vol 256,  © 2022 WIT Press
www.witpress.com, ISSN 1743-3541 (on-line)
doi:10.2495/ST220021

bioeconomy within the rest of the economy, not just advance it as a separate sector of interest to mainly rural communities. In particular, it is crucial to connect bioeconomy to the circular economy concept. Together they are stronger and make more sense in terms of reaching societal goals, than advancing them separately [2].

Contemporary tourism proves that experiencing different places is one of the main human interests, although this value today tends to be lost. In fact, modern man for a long time believed that science and technology had freed him from his direct dependence on places. Even in 1969, Lawrence Durell wrote: "As you slowly begin to learn about Europe, tasting its wines, cheeses and the characters of different countries, you begin to realize that the defining element of any culture is the spirit of place" [3]. Two-thirds of travellers will pay more money for products and services to companies that are committed to positive social and environmental impacts [4]. Tourist accommodation should reduce $CO_2$ emissions by 66% per room (2030) and 90% per room (2050) according to the Paris Agreement on Climate Change (2016). The U.N. World Tourism Organization predicts that by the year 2020, there will be some 1.6 billion eco-inspired trips taken. Eco-friendliness is evolving from a nice-to-have, on-trend hotel commodity to a must-have priority for a growing number of environmentally and socially conscious travellers [5].

The World Tourism Organization and the United Nations Development Programme "are committed to inspire leadership and facilitate collaboration to inspire stakeholders to advance the contribution of tourism to the SDGs". The year 2016 was a turning point: the 2030 Agenda for Sustainable Development and its sustainable development goals (SDGs) were adopted, and the Paris Agreement on climate change came into effect [6], [7]. These sent out a global political message on the way forward to transform our economic system to end poverty, protect the planet, and ensure prosperity for all. In this direction, while tourism is providing a significant boost to many local and national economies, mass tourism has been shown to pose a significant environmental defragmentation of the natural landscapes and fragile ecosystems. The hospitality industry uses substantial amount of energy and material resources while the embodied carbon produced in the whole supply chains seems crucial [8]. The effects on the natural environment include emissions to and pollutions off water resources, soil and the air, noise as well as the excessive use of locally available resources.

## 3 CASE STUDY: CAVES & WAVES

Caves & Waves is a sustainable hotel design manifesto that seeks to re-establish the "unconsciously repulsed" [9], [10] uprooting of man from nature: a statement that is vital to the current social and cultural reality of contemporary civilization, although it is a concept originally been implemented during the Romantic Revolution of the 19th century [11]. The architectural proposal, deeply rooted in the "Genius Loci" and the "phenomenology of architecture" [12], places the modern cosmopolitan "traveller – flaneur" [13] in the fundamental existential "in-between" space of the natural and the manmade place (topos). Thus, the proposal re-interprets the natural phenomena and the elements of the environment as a tool for the re-composition of the ground – terrain in "existential" terms in order to provide the infrastructure for man to "dwell" between the earth and the sky [14]–[16].

### 3.1  Interpreting the fragile ecosystem

The main concept of the architectural composition resulted from the lived experience of the place. A landscape characterised by sparse, mostly low, and bushy vegetation. Few trees – mostly cedars – are visible in the region. A harsh, dry, rocky, Mediterranean landscape with earthy colourations. A sloping landscape, eroded at its lowest point by constant direct contact with the sea.

### 3.1.1 Caves & Waves

Following an in-situ observation of the natural formations and cavities of the rocky terrain (CAVES) as formed by lengthy contact with the ocean vortices (WAVES), the central idea of the new hotel complex was composed, separated into two basic units. Firstly, the unit of habitation, as an archetypal "cave-womb" [17]–[19] – an intrauterine subterranean spatial quality – within the ground, and secondly, the horizontal surfaces of the "tamed" water that follows the ground elevation curves, as natural elements for the production of kinetic energy, capable of shaping both the dwelling-space, as well as the bioclimatic functionality of the hotel complex (Fig. 1).

Figure 1:    Bird's eye view – Building and landscape integration. *(Source: Alexandros Kitriniaris, KAAF (Kitriniaris Associates Architecture Firm), www.kaaf.gr. All rights reserved.)*

### 3.1.2 Inhabiting the womb

The first sense of unfamiliarity that modern man experiences when he comes to contact with nature occurs due to its "uprooting", its removal from it. Nature, which was completely familiar to primitive man, was alienated with the process of repulsion in the context of civilization and in-depth progress time and turned into something strangely unfamiliar (uncanny) to man. Therefore, modern man when it comes in contact with nature it experiences the unfamiliar as historically repulsed [20]. The concept of caved or dugout architecture first appears in nature from the beginnings of primitive habitation.

### 3.1.3 Return to nature

The lust of the endometrial existence, he mentions Freud, appears because the womb is the first abode that man experiences, without even being able to perceive it. His mother's womb is man's first refuge, in which he remains for nine months, and therefore the most intimate place he will ever experience. It is there that he feels for the first time the concept of protection and security. The connection to the womb refers to the need for man to return to where he came from, to the earth first and then symbolically inside the womb.

### 3.2 Sustainable design strategies

Waveforms express a characteristic structural determinism with their forms, depending on the reaction of the water to the forces acting through it. A vortex embodies the totality of dynamics in a balanced axis, which, after lengthy contact with the ground, shapes hollows that can potentially contain a residential space, as a "Vessel of Life" [21] capable of enveloping every primary human creative activity in a meaningful way. The archetypal space-womb creates a protected, subterranean hollow within which every human primary and creative activity develops both directly and meaningfully. The multiplication of the spatial womb comprises a structural complex that determines the primary residential space that is disengaged from personal peculiarities as it becomes incorporated into the natural landscape (see Fig. 2).

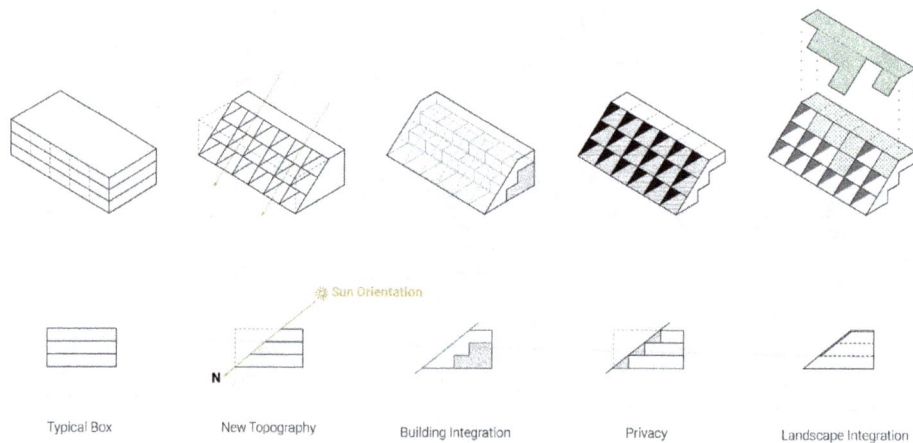

Figure 2: Main concept diagram. *(Source: Alexandros Kitriniaris, KAAF (Kitriniaris Associates Architecture Firm), www.kaaf.gr. All rights reserved.)*

### 3.2.1 Compositional principles
The building design comprises 2,500 m² of common public areas, 7,000 m² of private space (rooms), and 3,000 m² of auxiliary space. The compositional structure of the hotel comprises five clearly delineated zones (see Fig. 3). The circulatory and functional connection between the zones is achieved firstly through a smooth central network of internal circulation that follows the vertical curves and includes the movement of electric vehicles and pedestrians; and secondly, through a vertical subterranean system of wagonettes at the eastern border of the plot. The existing street planning system of the private settlement to the west forms the point of reference for carving each parallel zone into the landscape, while the basic levels of the complex's carved streets are articulated in parallel to the vertical curves of the terrain (see Fig. 4).

Figure 3:    Compositional exploded analysis diagrams. *(Source: Alexandros Kitriniaris, KAAF (Kitriniaris Associates Architecture Firm), www.kaaf.gr. All rights reserved.)*

### 3.2.2  Bioclimatic principles

The hotel complex integrates advanced energy and water utilization processes regarding its bioclimatic operation. The solar leaves which are placed in the oases of the swimming pools function on the one hand as a shelter and on the other hand collect and store the solar energy. Water reservoirs collect rainwater from the specially designed surfaces and roofs while reusing it either for internal use or for irrigation. The vegetation is mostly low similar to the surroundings while placing bigger trees on the rooms' landscaped areas to enhance the microclimate and protect the courtyards from the north winds. Finally, advanced actions have been taken to energy management of the hotel complex with the aim to cover operating costs over the project's lifetime and generate an acceptable rate of return (see Fig. 5).

### 3.2.3  Accessibility

The hotel complex is designed to be accessible for people with special needs. The accessibility of the caved buildings is done from corridors at the back side, in order to permit

Figure 4:    Masterplan. *(Source: Alexandros Kitriniaris, KAAF (Kitriniaris Associates Architecture Firm), www.kaaf.gr. All rights reserved.)*

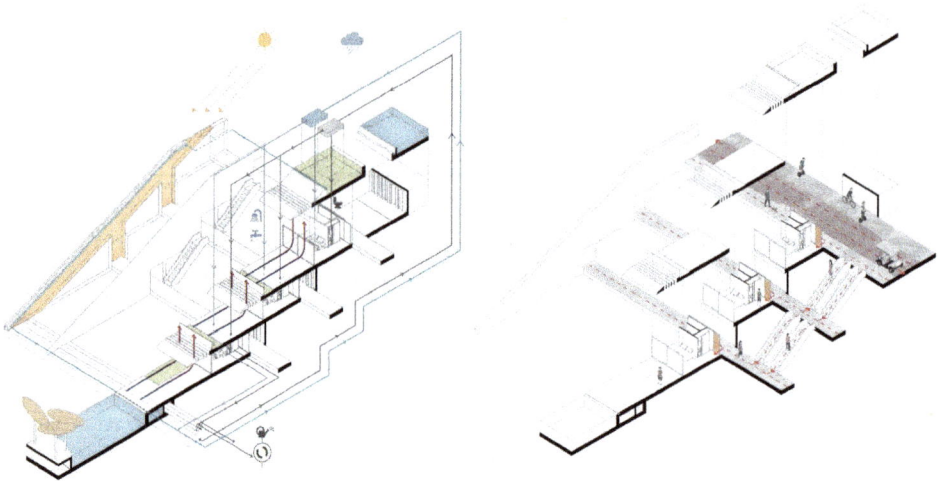

Figure 5:    Bioclimatic and accessibility diagrams. *(Source: Alexandros Kitriniaris, KAAF (Kitriniaris Associates Architecture Firm), www.kaaf.gr. All rights reserved.)*

natural lighting, cooling and ventilation for most of their length. In this way, the unobstructed view of all the rooms to the sea is always maintained without being interrupted. The rooms are accessed via escalators to ensure the two-way direction. Access to the rooms and common areas is ensured in every part of the complex while 20% of the rooms have direct access level from the internal road network.

### 3.2.4 Building and landscape integration

The phenomenon of repetition is understood as a process of the human brain, according to which the perception of an object is compared to other objects that are in close proximity to it. In this sense the synapses of the human brain compare only the neighbouring object to construct its structure while relating it to similar images. In this direction, the avoidance of the phenomenon of repeatability of the architectural design, is done through the gradual composition of the building units in the landscape in order to create irregular cavities for the penetration of the natural element (planting and water). In this way the perception of the building structure is perceived as fragments as it diffuses with the natural landscape (see Figs 6–10).

Figure 6: Exterior view of the dug-out units. *(Source: Alexandros Kitriniaris, KAAF (Kitriniaris Associates Architecture Firm), www.kaaf.gr. All rights reserved.)*

### 3.2.5 Material resources

The construction of the load-bearing structure of the caved buildings is made of reinforced concrete which is produced by cement companies that perfectly manage the $CO_2$ emissions and take care of the social responsibility of production of aggregates from reuse of raw materials. Reinforced concrete is also used for the construction of swimming pools and water tanks. Natural stone from the excavations is used as a cladding for the walls of reinforced concrete. Finally, natural wood is used as a material for ceilings, interior walls and floors, with the aim of reducing the ecological footprint. To this end, the project is constructed from natural and recyclable construction materials (local stone, wood), while contribute to global and local supply chains by minimizing embodied carbon of materials and promoting reusability and traceability [22] (see Fig. 11).

Figure 7: Exterior views of the waterfront and pool bar. *(Source: Alexandros Kitriniaris, KAAF (Kitriniaris Associates Architecture Firm), www.kaaf.gr. All rights reserved.)*

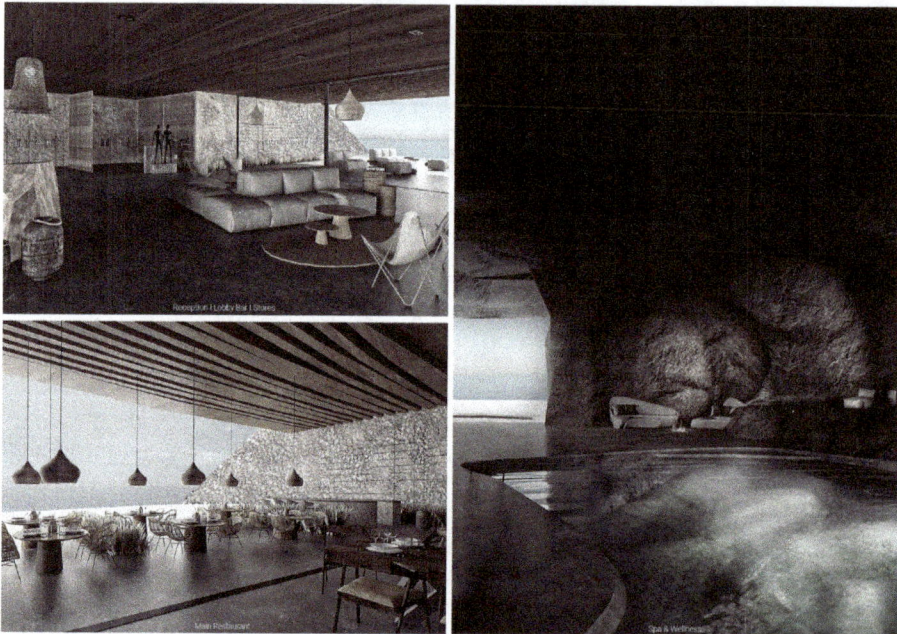

Figure 8: Interior view of the communal areas. *(Source: Alexandros Kitriniaris, KAAF (Kitriniaris Associates Architecture Firm), www.kaaf.gr. All rights reserved.)*

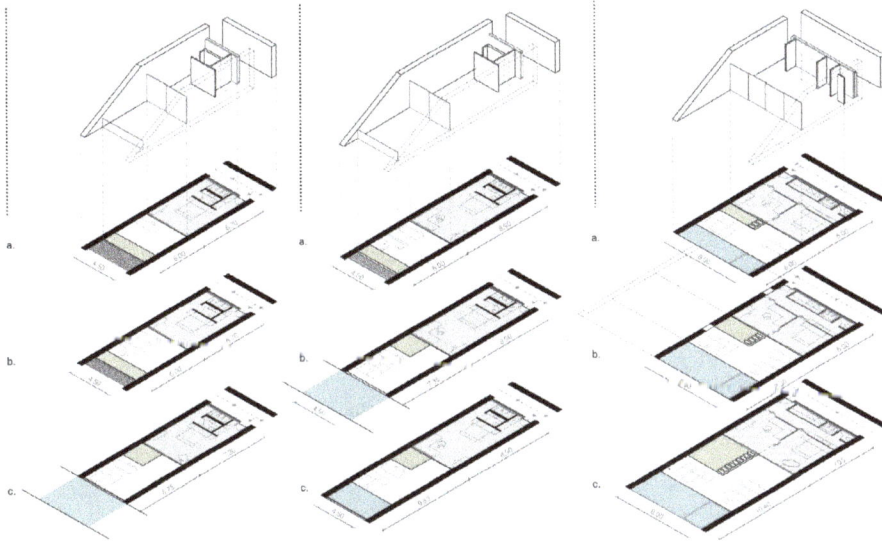

Figure 9:  Rooms and suites typology diagrams. *(Source: Alexandros Kitriniaris, KAAF (Kitriniaris Associates Architecture Firm), www.kaaf.gr. All rights reserved.)*

Figure 10:  Interior views of the rooms and suites. *(Source: Alexandros Kitriniaris, KAAF (Kitriniaris Associates Architecture Firm), www.kaaf.gr. All rights reserved.)*

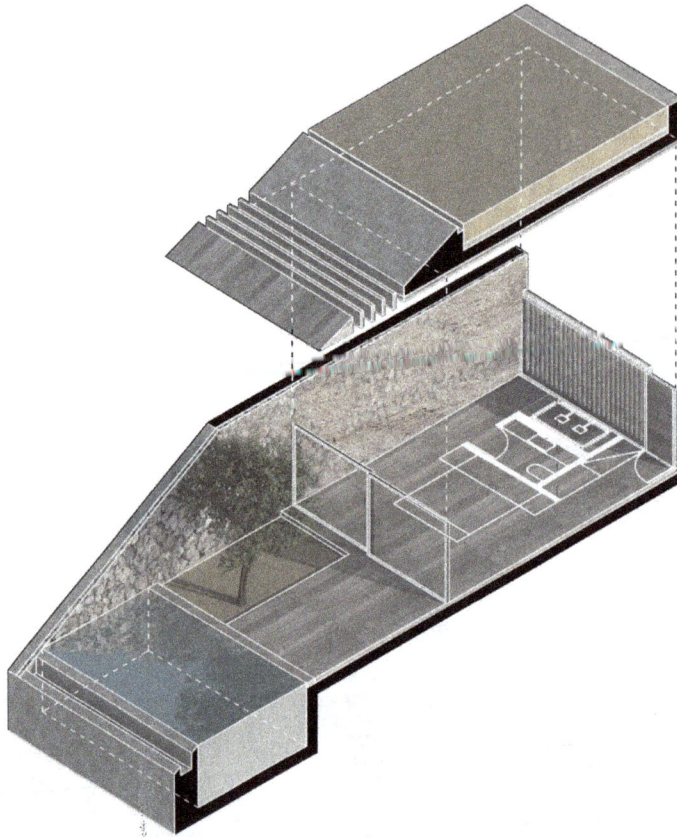

Figure 11:  Materiality diagram. *(Source: Alexandros Kitriniaris, KAAF (Kitriniaris Associates Architecture Firm), www.kaaf.gr. All rights reserved.)*

## 4 SUSTAINABILITY TARGETS OF HOTELS AND RESORTS

1. Sustainable hotels cover operating costs over their lifetime and generate an acceptable rate of return, while developing long-term monitoring methods to evaluate whether expectations and goals have been met.
2. Sustainable hotels demonstrate flexibility to adapt to future changes of user needs, ownership, laws, regulations, and economic fluctuations. In this case they create inventive programming strategies in terms of use, multiplicity of functions, short-term flexibility, and long-term adaptability.
3. Sustainable hotels contribute positively to reducing emissions and to the overall $CO_2$ balance of a structure's use-cycle. This means will be integrated to healthier environment and have advanced financial benefits from government funding, stakeholders, and investors.
4. Sustainable hotels minimizing their ecological footprint and maximizing their positive impact on the environment; by the reduction of harm and increase of beneficial effects. This means that hotels could categorized to green labels and increase the number of visitors who are keen on sustainable strategies, especially, visitors on Northern Europe.

5. Sustainable hotels, include innovative concepts regarding design, integration of materials and methods, structure, enclosure and mechanical systems.

6. Sustainable hotels improve existing contextual conditions responding to the natural and built environment. This means that create an added value to the surroundings, enhancing the local economy as well.

7. Sustainable hotels contribute to the formation of socially viable environments, strengthening of shared values and empowerment of communities, as well as adhere to ethical standards in all phases of the project.

8. Sustainable hotels, needs political transparency, unbiased processes and commitment to principled interaction, just practices, all in the effort to prevent corruption at every level. This includes the participation of stakeholders, including users, clients, neighbourhood affiliations, local authorities and non-governmental organizations.

9. Sustainable hotels create architectural quality and aesthetic impact, specifically concerning space, land, spatial sequences, movement, materials, light and ambiance.

10. In a nutshell, it is to respect the heritage of the region, to use local materials and work with simple and reliable systems with low environmental impacts. This is summarized under the following three points:
    - Zero kilometres: Nearness of construction materials and local skills
    - Zero carbon dioxide: Energy management and lower emissions.
    - Zero waste: Lifecycle management in the building process and reuse of building materials.

## REFERENCES

[1] Palahí, M. et al., Investing in nature to transform the post COVID-19 economy. *Solutions Journal*, **11**(2), 2020.

[2] Hetemäki, L., Hanewinkel, M., Muys, B., Ollikainen, M., Palahí, M. & Trasobares, A., Leading the way to a European circular bioeconomy strategy: From science to policy 5. European Forest Institute, 2017. DOI: 10.36333/fs05.

[3] Durell, L., *Spirit of Place: Letters and Essays on Travel*, Faber & Faber: London, 1969.

[4] Majumdar, B., How hospitality will become more sustainable in 2020. Multibriefs, 2020. https://exclusive.multibriefs.com/content/how-hospitality-will-become-more-sustainable-in-2020/travel-hospitality-event-management. Accessed on: 20 May 2022.

[5] World Tourism Organisation, https://www.unwto.org/en. Accessed on: May 2022.

[6] Sustainable Development Goals, https://sdgs.un.org/goals. Accessed on: May 2022.

[7] Paris Agreement, https://unfccc.int/process-and-meetings/the-paris-agreement/the-paris-agreement. Accessed on: May 2022.

[8] Bohdanowicz P. et al., Energy efficiency and conservation in hotels: Towards sustainable tourism. *4th International Symposium on Asia Pacific Architecture*, April 2001.

[9] Jentsch, E., *Zur Psychologie des Unheimlichen*, 1906.

[10] Freud, S., *The Uncanny*, translated by McLintock, D., Penguin Classics, 2003 (original 1919).

[11] Blanning, T., *The Romantic Revolution*, Phoenix, 2010.

[12] Norberg-Schulz, C., *Genius Loci, Towards a Phenomenology of Architecture*, Rizzoli: New York, 1980.

[13] Benjamin, W., *Charles Baudelaire: A Lyric Poet in the Era of High Capitalism*, Verso Books, 1997 (original 1969).

[14] Heidegger, M., *Sein und Zeit*, 1927. (Being and Time, translated by Joan Stambaugh, State University of New York Press, 1996).

[15]  Heidegger, M., *Essay on Building, Dwelling, Thinking*, 1951.
[16]  Heidegger, M., *The Thing*, 1950.
[17]  Vidler, A., *The Architectural Uncanny: Essays in the Modern Unhomely*, MIT Press, 1992.
[18]  Tzara, T., D'un certain automatisme du goût. *Minotaure*, **3–4**, 1933.
[19]  Kiesler, F.J., *Endless Space*, Hatje Cantz Publishers, 2001.
[20]  Freud, S., *The Uncanny*, Vol. XVII, The Hogarth Press: London, 1919.
[21]  Konstantinidis, A., Vessels for life or the problem of a genuine architecture. *Architecture Subjects*, Greece, 1972.
[22]  Designers and Specifiers, International Mass Timber Report, Chapter 5, Embodied Carbon and Biogenic Carbon, 2022.

# SECTION 2
# MARINE AND
# RURAL TOURISM

# COLD-WATER RECREATIONAL DIVING EXPERIENCES: THE CASE OF KELP FORESTS

SERENA LUCREZI & MICHAEL JUAN DU PLESSIS
TREES, North-West University, South Africa

## ABSTRACT

Recreational diving is an important part of marine tourism and has been extensively studied, especially in the context of warm-water and tropical environments. Comparatively, research on diving and the segmentation of divers in cold water habitats remains scant. Ever since the award-winning documentary My Octopus Teacher came out in 2020, more attention has been drawn to kelp forests, which are typically cold-water habitats. Diving in kelp is a recreational activity that has the potential to become a major part of marine tourism in countries like South Africa. However, limited research exists on this niche type of diving. This study aims to segment and profile kelp divers, to guide the sustainable development and management of kelp diving as marine tourism activity. To reach this aim, qualitative research was conducted. Specifically, semi-structured interviews were used to gather data on the demographic, psychographic, behavioural and specialisation profiles of 50 divers in Cape Town. The divers were male and female in similar proportion, they were in their late thirties and had tertiary education. They were mainly from South Africa and resided in Cape Town. They were both scuba and free divers and moderately specialised. Motivations to dive in kelp included observation, being in nature, relaxation and escape, adventure and exercise, discovery and learning, and photography. Experiences diving in kelp were relaxation and wellbeing, awe and wonder, contact with nature, freedom, novelty, feeling safe, and social interaction. The divers ascribed extrinsic and intrinsic values to kelp and recognised ecosystem services of kelp forests. They were also aware of problems affecting kelp and could propose viable solutions to combat these problems. The divers showed strong conservation commitment and place attachment to the local coast and kelp forests. The results were used to make recommendations concerning marketing and management, including better patrol of kelp forests, codes of conduct for divers, education, and marketing of sustainable kelp diving and other non-diving activities.

*Keywords: recreational diving, segmentation, recreation specialisation, diving motivation, phenomenology, values, attitude, conservation commitment, tourism management.*

## 1  INTRODUCTION

Recreational diving is an important and profitable branch of the marine tourism industry [1]. Diving can impact the areas surrounding dive sites, supporting coastal communities through income sources, job creation and overall economic growth in an area [2]. Diving also supports conservation directly and indirectly by creating awareness, knowledge of the sites and their ecological importance, and positive attitudes among divers towards the conservation of dive sites [3]. Much research has been done on the structure and impacts of recreational diving, to guide its management and sustainable development. There has been a focus on positive and negative environmental, economic, and social impacts [1]. Poor management of the activity can lead to problems between stakeholders making use of resources, habitats suffering from degradation, and revenue leakage from the local economy [4].

Understanding the structure and impacts of recreational diving first requires research on the segmentation of divers. Diving segmentation research has mainly focused on popular diving areas to enhance their management, including warm-water or tropical regions like Indonesia, Thailand, Malaysia, the Philippines, Australia, and the Red Sea [5], [6]. Recently, the focus has shifted to colder-water diving which is increasingly popular. Areas of study have included subtropical or temperate sites such as the Mediterranean Sea and South Africa

WIT Transactions on Ecology and the Environment, Vol 256, © 2022 WIT Press
www.witpress.com, ISSN 1743-3541 (on-line)
doi:10.2495/ST220031

[7], [8]. With increasing latitude, however, research on diving tourism remains proportionally scanty compared with lower latitudes, reflecting also the high niche activity nature of cold-water diving [9].

Kelp forests are seaweed forests that can be found in cold water on the coasts of temperate and Arctic zones [10]. These ecosystems house different aquatic life. They provide services to the life forms within them, such as food, habitat and protection against waves and bad weather, as well as services to humankind, such as fish stocks and recreation, in particular diving [11]. They are so biodiverse that they are compared to terrestrial rainforests and coral reefs in tropical destinations [12]. Given their compatibility with coral reefs, kelp forests potentially attract large numbers of divers and may require the same dedication towards tourism management and sustainability efforts as coral reefs do. This would call for research on kelp diving tourism, starting from segmentation. Studies have synthesised the economic value of kelp forests, including kelp tourism [13], [14]. The economic data from these studies suggest that kelp diving may be significant, and thus relevant to understand and monitor to ensure its management and sustainable development. However, limited or no research has looked at the state of the art of kelp tourism including diving. It would be important to begin by segmenting kelp diving, to manage the activity and secure its development as a sustainable recreational activity, cater for the different segments' needs, and prioritise actions to preserve kelp ecosystems and their services.

In South Africa, kelp distribution coincides with that of florid coastal tourism in Cape Town, a world-renowned coastal tourism destination surrounded by a marine protected area (MPA), where balancing human activities and conservation is paramount [15]. The local kelp forests were the setting of the award-winning Netflix documentary My Octopus Teacher. It has been argued that this and similar documentaries can raise awareness about ecosystems like kelp, possibly stimulating an interest in kelp tourism [16]. This documentary could draw attention to South Africa for kelp tourism including diving, which may represent an important factor of coastal and marine tourism in the area. Some research suggests this, although only indirectly [13], [15]. Thus, segmentation research on kelp diving is timely, given a research gap concerning this niche activity as well as cold-water recreational diving in general [9]. In times of COVID-19, inbound tourism has been subjected to a halt, which makes it important to explore the potential of domestic tourism and local coastal recreation. This study focused on recreational kelp diving by South African residents, contributing towards an appreciation and enhancement of domestic coastal tourism. A key question addressed in this paper is: what characterises segments of recreational kelp divers in South Africa? The study investigated the profile of kelp divers, with emphasis on their demographic profile, diving specialisation, motivations, the phenomenology of kelp diving, and conservation commitment. The novelty of this research is grounded on addressing a research gap in the segmentation of kelp divers, as well as contributing to segmentation research on cold-water diving tourism.

## 2 METHOD

This study used qualitative research and a phenomenology approach. According to Sokolowski [17], phenomenology is "the study of human experiences and of the way things presents themselves to us through such experiences". A semi-structured telephonic interview was used to collect the data. The interview was structured to reflect different constructs in diving segmentation research, including demography, specialisation, motivations, and experiences (phenomenology) [18], [19]. Questions revolving around perceptions of kelp, including values and attitude, as well as conservation commitment, were added to understand the relationship between diving segments and disposition towards marine environments and kelp [20], to better inform management, especially considering that kelp forests in Cape Town fall within the boundaries of an MPA.

Interviews were conducted and audio-recorded by a field worker with the consent of the participants. Once a week in September to December 2020, the fieldworker posted a message on randomly selected diving groups on social media. These included two WhatsApp groups of recreational divers in Cape Town, and two Facebook groups representing diving schools and shops in Cape Town. The message invited participants to partake in a half-hour interview via telephone (using WhatsApp or Messenger) to discuss diving in kelp. The message was formulated neutrally to attract different types of divers and a date was set by divers in which they could be interviewed. After 50 divers were interviewed, the data were transcribed to check for saturation, which had been achieved.

The recordings were transcribed verbatim into Microsoft Excel. Numerical and categorical answers, such as demographic ones, were analysed using basic statistics including frequency tables and averages. The transcribed data were analysed using thematic analysis, which is a method that is used in research to identify, sort and get meaning from patterns or themes that are found when doing qualitative research [21].

## 3 RESULTS AND DISCUSSION

### 3.1 Demographic profile and diving specialisation

Just over half (56%) of the participants were female as opposed to 44% male. Most were in their late thirties and had tertiary education. The participants mainly came from South Africa and resided in Cape Town. Only three participants claimed to be in a marine profession. The demographic profile of the participants partly corresponds with what is found in literature concerning divers, with a focus on the cold-water market. For example, the age range and education level of the participants matched with those highlighted by other authors [22], [23]. More females than males participated in the study, which is at odds with statistics from diving agencies like PADI [24]. Since the sample was characterised by many free divers, this group may have a high female representation, although no statistics could be found in this regard. It is also possible that females were more interested in participating in this study; research has shown that females are more cooperative and compliant when it comes to social science surveys [25]. The origin of the participants suggests that kelp divers in South Africa (it must be considered that this study happened during the closure of the country to international tourism) are local and from the Cape. Kelp forests can only be found in the cold-water regions of South Africa and around the Cape Peninsula [13], implying that the divers dived their local coast. This result also suggests attachment to the local coast, encountered in other cold water diving research [22].

Most participants practised both free diving and scuba diving and had a basic (e.g. PADI Open Water Diver) or advanced (e.g. PADI Advanced Free Diver) diving qualification. Among the participants, an average of 819 lifetime dives had been logged. Most had been diving for over 10 years and logged just over 65 dives per year. They were mainly not employed in the diving industry and were not subscribed to a dive magazine but belonged to a diving club. Cold-water divers are ranked high on the specialisation continuum, which goes from generalist or low-commitment to specialist or high-commitment people [7], [22], and the data in this study confirm this. Not many participants were employed in the diving industry and this result may be explained in their motivations to dive (discussed later), which included elements of escapism. The participants were part of a dive club which indicates that they are serious about the sport and want to be part of a community that is like-minded – this is also an indication of a commitment to the sport and specialisation [26]. Being subscribed to a diving magazine may not be popular anymore since information about diving is now digital and can be found online [27].

## 3.2  Motivations to dive in kelp

Motivations to dive in kelp included observation, being in nature, relaxation and escape, adventure and exercise, discovery and learning, and photography (Table 1). Observation and being in nature were mentioned in the largest proportion (half of the participants). Observation focused on seeing the creatures thriving in the kelp forest as well as the aesthetic qualities of kelp. Being in nature involved wanting to be fully immersed in the kelp forests. Divers compared the aesthetic qualities of kelp to those of rainforests or described kelp diving as walking in a botanical garden. They also wanted to feel as if they were flying through a forest, which was described as a magical experience. Another two frequently mentioned motivations (over 30% of divers) were relaxation and escape, and adventure and exercise. The former is about escaping from life and finding a place where to relax. Kelp forests provided divers with a place where they could feel safe and where the world and everyday worries would not bother them. Diving in kelp was also a form of adventure, where divers could have "out of this world" experiences, be challenged, and be able to exercise and stay fit. Motivations that were described in lesser proportion (around 10%) included discovery and learning, and photography. Some people dived in kelp because they had an interest in learning while being engaged in a recreational activity. These divers wanted a place where they could learn about biodiversity and the organisms that can be found. Divers also enjoyed the opportunity to take pictures of the kelp forest and the creatures within. The underwater landscapes as well as documenting underwater life for further research were important driving factors in this case.

Motivations partly match with what is described regarding cold-water diving, for example, adventure, learning, and escapism [22], [23]. Motivations themes that emerged more strongly in kelp diving compared with general cold-water diving included observation, and the desire to dive in kelp to have magical experiences in nature. While observation is a more typical motivation in diving tourism, where curiosity is a strong driver, being in nature is also recurrent in marine ecotourism, adventure tourism, and nature sports literature [1], [28]–[30]. The results suggest that on the one hand, kelp diving may offer experiences that are similar to those offered in other types of cold-water diving, such as adventure and learning. On the other hand, it has unique properties attracting curious people wanting to enjoy the aesthetic qualities of the dive site and to be fully immersed in nature.

## 3.3  Experience of diving in the kelp

Experiences or phenomenology of kelp diving included relaxation and wellbeing, awe and wonder, contact with nature, freedom, novelty, feeling safe, and social interaction (Table 1). Relaxation and wellbeing were mentioned by most (62%). Divers found that diving in kelp is a meditative experience and a way to relax and step away from the stresses of everyday life. They declared that kelp diving helped them connect with their spiritual side as well as the beliefs they hold. It helped them "slow down" and "be in the moment", and was also considered a coping mechanism when things got too difficult in life. It was an activity contributing to mental and physical wellbeing. The divers mentioned the benefits of cold water for human physiology, such as helping blood pressure. Other sub-themes frequently

Table 1: Main sub-themes extracted from the narratives of participants in this study (N = 50), and key quotes for each sub-theme.

| Sub-theme | Key quote |
| --- | --- |
| **Motivations to dive in kelp** | |
| Observation | I want to see the life there, it is so interesting and beautiful. |
| Being in nature | I want to experience nature and witness local ecosystems. |
| Relaxation and escape | Free diving is total and utter relaxation. |
| Adventure and exercise | There is a lot of life and adventure to be found. |
| Discovery and learning | I dive for the colour, fish, critters, rocks, and the ecosystem where you find beautiful creatures that you would not find elsewhere. |
| Photography | The scenery for photography can be spectacular with good visibility. |
| **Experience of diving in the kelp** | |
| Relaxation and wellbeing | The feeling is calm and relaxed, loving every second. It is very good for my mental health and wellbeing to go into the kelp forest. |
| Awe and wonder | I just have feelings of awe and wow when I am in the kelp forest. |
| Contact with nature | I learned that kelp is your friend; it makes me feel at home to dive in the kelp. |
| Freedom | I go without wetsuits and gloves, I find it separating if you do. |
| Novelty | The first time I had anxiety, but then at the bottom, it was a lot quieter and got used to it. |
| Feeling safe | I get over my fear, it is about self-mastery. |
| Social interaction | I was comfortable around my friends and instructors. |
| **Perceived value of kelp** | |
| Ocean ecosystems | Like rainforests, kelp creates an environment supporting life. |
| Marine life | A huge food source and many species lay their eggs on the kelp. |
| People | Kelp has psychological benefits for people suffering from stress. Cold water and waves have calming effects. Without mentioning the massive tourism value. People coming to see great whites and seals, it is all connected. You will not have opportunities to snorkel, free or scuba or fish without kelp here. |
| **Perceived issues/solutions for kelp conservation** | |
| Problems affecting kelp | Maybe climate change: kelp prefers colder temperatures, changing water temperatures and currents may affect distribution. Human impact (e.g. overfishing) causes an imbalance in the food chain. Removing the predators of animals that eat kelp, you get these barren lands. |
| Ways to protect kelp | We could designate protected and controlled reserves, but they need to be effective and we should create networks of interconnected sites, species are mobile. This may be important for climate change because species are likely to migrate. Also regulate activities within the kelp forests, whether it is recreation or fishing. |

mentioned were awe and wonder (44%) and being in nature (40%). For divers, it was a wonder to dive in kelp and an experience that humbled them. They were mesmerised by the movement of the kelp forests, which changed their perspectives on life and the "surface world". Divers claimed that they had a deeper connection with nature when diving in the kelp forests, and enjoyed having interactions with nature and witnessing the balance of life. The connection with nature created awareness of the fragility of marine ecosystems.

Sub-themes mentioned by almost a third of the divers included freedom (26%) and novelty (22%). The former was related to the sense of freedom derived mainly from free diving as opposed to scuba diving, and from diving without basic equipment like wetsuits or gloves to feel in better contact with the water element. Novelty revolved around first-time experiences for some of the participants. For example, diving in kelp brought new experiences to already professional divers. Some of the divers stated how they were scared at first and felt claustrophobic but later came to enjoy kelp diving. Other divers added that kelp diving allowed them to switch from scuba to free diving to avoid incidents like being stuck in kelp. Feeling safe was an experience mentioned by 14%. Here, the kelp was described as a place protecting divers from the larger predators of the ocean while they had the opportunity to enjoy the scenery. For some, the experience was a combination of scary and feeling safe while for others, it was a way to conquer their fears. The last sub-theme was social interaction (6%), which revolved around experiences with family and friends during kelp diving. This was not a frequently mentioned sub-theme, because kelp diving emerged as an activity that divers enjoyed doing alone despite being connected with like-minded people.

The phenomenology of kelp diving echoed relevant sub-themes in the motivations category, especially relaxation and wellbeing, awe and wonder, and contact with nature. Kelp diving was described as a highly satisfactory experience stimulating the senses, generating mental and physical benefits, and creating awareness, in line with similar ecotourism activities [31]. The view that kelp diving is an invigorating and therapeutic activity echoes research focusing on the beneficial role of cold water in tourism [32]. Sinclair [33] confirmed the critical role of cold-water experiences as part of blue exercise on mental and physical wellbeing and their positive relationship with a sense of spirituality (including relaxation, meditation, awe and wonder, and contact with nature). Elmahdy [34] added that cold-water activities, such as cold-water surfing, effectively contribute to benefits including wellbeing, a deeper connection with nature, and feelings of freedom. The results highlighted a "sensorium seeking" or the involvement of the senses in the experience, which included kinaesthesia [35].

The participants experienced diving in kelp as heightening the senses and their awareness while making them feel like losing themselves, their social stature, and track of time, plus a sense of personal challenge especially in the case of free divers. These elements resemble "flow states" which have emerged in previous literature related not only to diving but also other nature-based recreational activities and sports [36], [37]. The divers described their immersion in cold water and the kelp forests as a unique experience, characterised also by interactions with the ecosystem and the silent observation of the unfolding of nature. This experience was also epiphanic since it led the divers to feel small and meditate on the greatness and vulnerability of nature. These experiences are typical of memorable embodiment experiences in wildlife tourism (and not exclusively diving) [38]. The divers' narratives often revolved around personal and environmental contexts which contribute to place attachment, with emphasis on place dependence, place identity, nature bonding and environment bonding [39].

### 3.4 Extrinsic and intrinsic values ascribed to kelp

The perceived role and importance of kelp for the functioning of ocean ecosystems, marine life and people included different ecosystem services (provisioning, supporting, regulating, and cultural) of kelp (Table 1). The divers mentioned ecosystem engineering (i.e. species shaping and modifying ecosystems), coastal protection, water filtration, carbon sequestration, photosynthesis and oxygen production, habitat creation, food provision, wildlife protection (e.g. shelter, nursery grounds), raw materials for industries (e.g. health, agriculture), and tourism.

The results of the study show that kelp divers can identify extrinsic and intrinsic ecosystem services of kelp, with a strong focus on supporting ones but also mentioning regulatory, provisioning and cultural ones [10], [11], [13]. The participants identified and described ecosystem services that benefit people, but also emphasised the value of kelp for the ocean and species, such as biodiversity [11]. They did not differentiate between the natural environment and humankind, recognising that everything is connected. The cultural benefits ascribed to kelp echoed the most important experience theme identified, namely relaxation and wellbeing.

### 3.5 Problems affecting or threatening kelp and ways of mitigation

With regards to problems threatening kelp, most participants were concerned about pollution and global warming (Table 1). Other threats that were identified included the poaching and overfishing of certain species that help to keep kelp under control (removal of abalone causing a rise in sea urchins feasting on kelp). Other impacts that divers felt were important were coastal urbanisation as well natural phenomena (such as storms) and direct impacts such as from boats and divers. Concerning solutions to kelp threats, the divers mentioned no-take or no-go zones and the broadening of existing MPAs to create wider networks. Other important ways to combat kelp threats were to implement stricter regulations with regards to fishing and harvesting of kelp as well as getting the government involved with combating threats. The reduction of carbon emissions was also one way most of the participants indicated kelp problems could be stopped. Education and raising public awareness were also mentioned as ways to promote the protection of kelp, indirectly.

These results confirm that kelp divers are aware of several problems affecting kelp ecosystems, from pollution to overfishing and direct damage, as well as efforts to combat these problems [11], [12]. Divers were concerned about the environment in which they practised their activities, in line with research on the behavioural segmentation of divers, specialised divers and cold-water divers. Diving specialisation can be positively related to greater knowledge of and attachment to dive sites [18], [26]. Cold-water divers can show high levels of concern for and strong positive attitudes towards the environment [7], [22]. The results also show that kelp divers are strongly attached to the kelp forests of Cape Town and know what local problems are affecting kelp as well as ways to approach these problems [13].

### 3.6 Participation in marine conservation projects

Most participants (58%) had previously contributed to marine conservation projects. These included clean-up, research, film and photography, citizen science, and miscellaneous conservation projects (e.g. volunteering for bird conservation). Participation was with organisations that hosted conservation activities (e.g. large organisations monitoring coral, aquariums partaking in wildlife rehabilitation). All participants were willing to be a part of kelp conservation projects in the future. These results show a strong conservation commitment on behalf of the participants in this study. Other research has demonstrated that cold-water divers show good conservation behaviour and keep up to date with conservation documentaries [7], [23], [26]. What was new with kelp divers was their determination about getting the community and the public involved with kelp conservation projects. Many said that the public needs to be educated and their knowledge and awareness about kelp needs to be increased.

## 4 CONCLUSION AND RECOMMENDATIONS

The study contributed towards filling a gap in knowledge about cold-water tourism in South Africa, with a focus on kelp diving. Given the circumstances of the global COVID-19 pandemic, it was only possible to investigate domestic recreation revolving around kelp forests. However, this factor provided an opportunity to promote new niches and guide their sustainable development. Based on the results of this study, the following management and marketing recommendations could be made.

### 4.1 Management

While the results of the study show that there should be no concern regarding the characteristics and behaviour of current kelp divers, the development and management of kelp tourism should remain sustainable. Documentaries like My Octopus Teacher might increase the demand for kelp tourism and this would need to be channelled into the proper promotion of this type of tourism, including diving. Since most of the kelp diving in South Africa happens within MPAs, the management plan of these areas should include regulations specific to kelp diving and kelp tourism in general (e.g. snorkelling, kayaking). Since South African MPAs lack sufficient manpower to enforce regulations [40], a possibility to enhance control of kelp tourism would be to have dive and tourism operators design and implement codes of conduct and best practice. These can include rules such as no-touch and no harassment of animals when diving. Additionally, dive and tourism operators have the potential to educate current and new kelp tourists including divers, acting as mediators and mitigators of improper conduct. Education surrounding kelp and codes of conduct and best practice can take place at dive centres, for example through briefings and de-briefings. Dive centres often host marine biologists and special guest speakers who talk about ecology and environmental threats. Kelp could feature as one of the themes discussed during planned events with these people, especially in areas like Cape Town where kelp forests are a dominant seascape.

Another management approach would be to distribute (due to a lack of resources) some of the responsibilities of MPA management bodies to supporting entities like NGOs and have different stakeholders working together to enhance patrol of MPAs and enforcement of regulations and codes of conduct around ecosystems including kelp forests. NGO and local tour operations and dive centres could be in charge of the codes of conduct and best practices at the MPAs as well as monitoring kelp tourism, including kelp divers and their behaviour. These role-players could have ambassadors and wardens to monitor kelp forests and guide education in and around kelp forests. For example, selected diving personalities, such as famous photographers or biologists, could be elected "kelp champions" and advocate for proper kelp tourism and diving using their communication channels (e.g. social networks, public talks).

## 4.2 Marketing

The results of this study show that the current kelp diving market in South Africa possesses certain characteristics that can be useful to direct future marketing of kelp diving as a form of tourism, as well as other forms of kelp tourism. Kelp diving can be marketed as a form of ecotourism. This is because it is an activity with low environmental impact and practised by conscious people who are committed to conservation and wish to observe and learn about kelp ecosystems. New offers can be created to cater for the needs of the current market of kelp divers, based on the characteristics that they share. For example, divers could be interested in offers including educational components, interpretation and participatory research or citizen science (e.g. monitoring kelp). Kelp tourism can be promoted to potential tourists, including both divers and non-divers. For example, dive centres and dive training facilities can use kelp forests as the perfect setting to teach people to dive, especially free diving. As for divers who have never dived in kelp, the same could be applied, especially if divers can scuba but do not free dive (free diving is easier than scuba around the kelp, as also highlighted in the narratives of the participants in this study). For scuba divers who do not wish to free dive, considering that kelp diving is a form of cold-water diving, dive centres could offer drysuit courses to enable access to cold water.

The participants in this study felt that diving in kelp was like floating in a forest and the aesthetic qualities of the kelp were an important element of the diving experience. Therefore, kelp forests can represent perfect settings for underwater photography and portrait photoshoots, creating products that can be attractive to divers, photographers and artists. The study also showed the strong meditative and therapeutic benefits of kelp diving. One could take advantage of this by promoting activities that can harness these benefits. For example, "kelp yoga" sessions can be offered by trained professionals to divers; these sessions can take place on the beach before diving sessions. Considering that kelp is abundant and a prominent feature of the coast in the Cape Peninsula, kelp could represent a symbol of the national heritage, to be promoted through events that can attract tourism and contribute to the concept of "Brand South Africa". South Africa already is popular for creating festivals around marine life such as the annual Oyster Festival and the Whale Festival [41]. The City of Cape Town, in collaboration with local tour operators, businesses, communities and organisations, can organise and promote a kelp festival where tourists can experience all attributes of kelp (e.g. kelp as food, kelp as art) without having to dive. The idea of a kelp festival can contribute to the growing trend of slow tourism in and around Cape Town which has become popular in the past few years [42].

## 4.3 Conclusion

This study segmented and profiled kelp divers in South Africa with a focus on motivations, experiences and conservation commitment. The results informed the management and marketing of kelp diving tourism in the country as well as other potential forms of kelp tourism. There are limitations to this study that need to be taken into consideration in interpreting the findings and planning future research. COVID-19 caused some problems such as limited access to other markets and potential participants, particularly international visitors. The sample was relatively small with only 50 divers interviewed, although data saturation was reached. The number of variables investigated was also limited. Nevertheless, the study contributed important information with regards to domestic recreation during times of COVID-19, which are historically significant and will define the future of tourism trends.

## ACKNOWLEDGEMENTS

Special thanks go to the participating divers, Terry Corr, Carlo Cerrano and Antonietta d'Agnessa. This work was funded by the National Research Foundation of South Africa (NRF Grant Number 119923), which was not involved in any part of the research. The research was approved by the Research Ethics Committee of the Faculty of Economic and Management Sciences at the North-West University, South Africa, under the ethics code NWU-00853-20-A4. This paper reflects only the authors' view. The NRF and North-West University accept no liability whatsoever in this regard.

## REFERENCES

[1]   Musa, G. & Dimmock, K., Introduction: Scuba diving tourism. *SCUBA Diving Tourism*, eds D. Musa & K. Dimmock, Routledge: London, pp. 21–31, 2013.

[2]   Arcos-Aguilar, R., Favoretto, F., Kumagai, J.A., Jimenez-Esquivel, V., Martinez-Cruz, A.L. & Aburto-Oropeza, J., Diving tourism in Mexico: Economic and conservation impotence. *Marine Policy*, 2021. DOI: 10.1016/j.marpol.2021.104410.

[3]   Angulo-Valdes, J.A. & Hatcher, B.G., A new typology of benefits derived from marine protected areas. *Marine Policy*, **34**(3), pp. 635–644, 2010.

[4]   Dimmock, K. & Musa, G., Scuba diving tourism system: A framework for collaborative management and sustainability. *Marine Policy*, **54**, pp. 52–58, 2015.

[5]   Edney, J. & Boyd, W.E., Diving under the radar: Divers and submerged aircraft. *Journal of Heritage Tourism*, **16**(1), pp. 100–117, 2021.

[6]   Lucrezi, S., Ferretti, E., Milanese, M., Sara, A. & Palma, M., Securing sustainable tourism in marine protected areas: Lessons from an assessment of scuba divers' underwater behaviour in non-tropical environments. *Journal of Ecotourism*, **20**(2), pp. 165–188, 2021.

[7]   Lucrezi, S., Milanese, M., Sarà, A., Palma, M., Saayman, M. & Cerrano, C., Profiling scuba divers to assess their potential for the management of temperate marine protected areas: A conceptual model. *Tourism in Marine Environments*, **13**(2–3), pp. 85–108, 2018.

[8]   Lucrezi, S., Bargnesi, F. & Burman, F., "I would die to see one": A study to evaluate safety knowledge, attitude, and behaviour among shark scuba divers. *Tourism in Marine Environments*, **15**(3–4), pp. 127–158, 2020.

[9]   Lew, A.A., A world geography of recreational scuba diving. *SCUBA Diving Tourism*, eds D. Musa & K. Dimmock, Routledge: London, pp. 29–51, 2013.

[10]  Teagle, H., Hawkins, S.J., Moore, P.J. & Smale, D.A., The role of kelp species as biogenetic habitat formers in coastal marine ecosystems. *Journal of Experimental Marine Biology and Ecology*, **492**, pp. 81–98, 2017.

[11]  Wernberg, T., Krumhansl, K., Filbee-Dexter, K. & Pedersen, M.F., Status and trends for the world's kelp forests. *World Seas: An Environmental Evaluation*, ed. C. Sheppard, Academic Press: Cambridge, pp. 57–78, 2019.

[12]  Bennett, A., Wernberg, T., Connell, S.D., Hobday, A.J., Johnson, C.R. & Poloczanska, E.S., The "great southern reef": Social ecological and economic value of Australia's neglected kelp forests. *Marine and Freshwater Research*, **67**(1), pp. 47–56, 2016.

[13]  Blamey, L.K. & Bolton, J.J., The economic value of South African kelp forests and temperate reefs: Past, present and future. *Journal of Marine Systems*, **188**, pp. 172–181, 2018.

[14]  Merkel, A., Säwe, F. & Fredriksson, C., The seaweed experience: Exploring the potential and value of a marine resource. *Scandinavian Journal of Hospitality and Tourism*, 2021. DOI: 10.1080/15022250.2021.1879671.

[15] Pfaff, M.C. et al., A synthesis of three decades of socio-ecological change in False Bay, South Africa: Setting the scene for multidisciplinary research and management. *Elementa: Science of the Anthropocene*, **32**(7), 2019. DOI: 10.1525/elementa.367.

[16] Reid, V., My octopus teacher. *Biodiversity*, **21**(3), p. 168, 2020.

[17] Sokolowski, R., *Introduction to Phenomenology*, Cambridge University Press: New York, 238 pp., 2000.

[18] Cater, C., Albayrak, T., Caber, M. & Taylor, S., Flow, satisfaction and storytelling: A causal relationship? Evidence from scuba diving in Turkey. *Current Issues in Tourism*, **24**(12), pp. 1749–1767, 2020.

[19] Lucrezi, S., Saayman, M. & Van Der Merwe, P., Managing diving impacts on reef ecosystems: Analysis of putative influences of motivations, marine life preferences and experience on divers' environmental perceptions. *Ocean and Coastal Management*, **76**, pp. 52–63, 2013.

[20] Gkargkavouzi, A., Paraskevopoulos, S. & Matsiori, S., Public perceptions of the marine environment and behavioral intentions to preserve it: The case of three coastal cities in Greece. *Marine Policy*, **111**, 103727, 2020. DOI: 10.1016/j.marpol.2019.103727.

[21] Braun, V., Clarke, V. & Terry, G., Thematic analysis. *Qualitative Research on Clinical and Health Psychology*, eds P. Rohleder & A.C. Lyons, Palgrave Macmillan: New York, pp. 95–113, 2014.

[22] Hermoso, M.I., Martin, V.Y., Gelcich, S., Stotz, W. & Thiel, M., Exploring diversity and engagement of divers in citizen science: Insights for marine management and conservation. *Marine Policy*, **124**, 104316, 2021. DOI: 10.1016/j.marpol.2020.104316.

[23] Thapa, B., Graefe, A.R. & Meyer, L.A., Moderator and mediator effects of scuba diving specialization on marine-based environmental knowledge-behavior contingency. *The Journal of Environmental Education*, **37**(1), pp. 53–67, 2005.

[24] Professional Association of Diving Instructors (PADI), 2019 Worldwide corporate statistics. Data for 2013–2018. https://www.padi.com/sites/default/files/documents/2019-02/2019%20PADI%20Worldwide%20Statistics.pdf. Accessed on: 10 Mar. 2021.

[25] Fife-Schaw, C., Surveys and sampling issues. *Research Methods in Psychology*, **2**, pp. 88–104, 2000.

[26] Lucrezi, S., Milanese, M., Cerrano, C. & Palma, M., The influence of scuba diving experience on divers' perceptions, and its implications for managing diving destinations. *PloS One*, **14**(7), 2019. DOI: 10.1371/journal.pone.0219306.

[27] Scholtz, M. & Kruger, M., From drifters to followers: A CIA-typology for engaging followers of scuba dive operators' Facebook pages. *Current Issues in Tourism*, **23**(18), pp. 2283–2301, 2020.

[28] Houge Mackenzie, S. & Brymer, E., Conceptualizing adventurous nature sport: A positive psychology perspective. *Annals of Leisure Research*, **23**(1), pp. 79–91, 2020.

[29] Junot, A., Paquet, Y. & Martin-Krumm, C., Passion for outdoor activities and environmental behaviors: A look at emotions related to passionate activities. *Journal of Environmental Psychology*, **53**, pp. 177–184, 2017.

[30] Melo, R. & Gomes, R., Nature sports participation: Understanding demand, practice profile, motivations and constraints. *European Journal of Tourism Research*, **16**, pp. 108–135, 2017.

[31]  Hansen, A.S., The visitor: Connecting health, wellbeing and the natural environment. *Tourism, Health, Wellbeing and Protected Areas*, ed. I. Azara, CABI: London, pp. 125–137, 2018.

[32]  Brito Henriques, E., Sarmento, J. & Lousada, M.A., When water meets tourism: An introduction. *Water and Tourism: Resources Management, Planning and Sustainability*, Lisbon, Portugal, Centro de Estudos Geográficos, Universidade de Lisboa, pp. 13–33, 2010.

[33]  Sinclair, J., The wave model: A holistic exploration of the sea's positive effect on wellbeing. Presented at *5th Annual Applied Positive Psychology Symposium*, Buckinghamshire New University, High Wycombe, UK, 2019.

[34]  Elmahdy, Y.M., "100 percent fun": A case study of benefits from cold water surfing in Jæren, Norway. University of Stavanger, Norway, 2015.

[35]  Allen-Collinson, J. & Hockey, J., Feeling the way: Notes toward a haptic phenomenology of distance running and scuba diving. *International Review for the Sociology of Sport*, **46**(3), pp. 330–345, 2011.

[36]  Dimmock, K., Comfort in adventure: The role of comfort, constraints and negotiation in recreational scuba diving. Southern Cross University, Lismore, 2009.

[37]  Nakamura, J. & Csikszentmihalyi, M., The concept of flow. *Flow and the Foundations of Positive Psychology*, ed. M. Csikszentmihalyi, Springer: London, pp. 239–263, 2014.

[38]  Curtin, S., What makes for memorable wildlife encounters? Revelations from "serious" wildlife tourists. *Journal of Ecotourism*, **9**(2), pp. 149–168, 2010.

[39]  Haywood, B.K., Parrish, J.K. & He, Y., Shapeshifting attachment: Exploring multi-dimensional people–place bonds in place-based citizen science. *People and Nature*, **3**(1), pp. 51–65, 2021.

[40]  Chadwick, P., Duncan, J. & Tunley, K., State of management of South Africa's marine protected areas. WWF South Africa Report Series, 1–209. http://awsassets.wwf.org.za/downloads/mpastateofmanagementreport04nov2009webl owerdpi.pdf. Accessed on: 10 Mar. 2021.

[41]  Kruger, M., van der Merwe, P. & Saayman, M., A whale of a time! An experience-based typology of visitors to a South African whale-watching festival. *Journal of Outdoor Recreation and Tourism*, **24**, pp. 35–44, 2018.

[42]  Lowry, L.L. & Back, R.M., Slow food, slow tourism and sustainable practices: A conceptual model. *Sustainability, Social Responsibility and Innovation in Hospitality-Tourism*, eds H.J. Parsa & V. Narapareddy, CRC Press: Oakville, Canada, pp. 71–89, 2015.

# FISHING TOURISM AND SUSTAINABILITY
# IN THE CANARY ISLANDS, SPAIN

PABLO DÍAZ RODRÍGUEZ & ALBERTO JONAY RODRÍGUEZ DARIAS
University Institute of Social Research and Tourism, La Laguna University, Spain

## ABSTRACT

Fishing tourism is an activity that has shown great potential to improve the living conditions of fishing populations, reduce the pressure on fishery resources and singularize the tourism offer of coastal tourist destinations. However, its adequate formulation requires strategies for the participation of the different groups of agents involved in the activity. In addition, the incorporation of fishermen as tourism service providers would require the development of training and awareness strategies. From an ethnographic approach, this paper analyses the possibilities and limitations for the development of fishing tourism in the Canary Islands, after the approval of Law 15/2019 of 2 May of the Canary Islands Fishing, showing the perceptions of professional fishermen about the barriers and potentialities of the incorporation of traditional fishing to the list of tourist experiences in the Canary Islands.
*Keywords: massive sun and beach destinations, Canary Islands, fishing tourism, regulation, sustainable development.*

## 1 INTRODUCTION

A highly significant portion of the large mass tourism destinations that have been developed internationally since the 1960s are located in coastal areas with favourable climatic conditions. However, tourism activities are not normally the only human activities that take place in these scenarios. Fishing and shell fishing have coexisted for decades with the massive tourist exploitation of the coast in multiple destinations and contexts.

The relationship between coastal mass tourism and fishing activities is very diverse. On the one hand, tourism is a source of business for fishing populations due to the increased demand for fishery products, the emergence of new job opportunities, or the possibility of participation in tourism entrepreneurship or through tourism marketing of fishing activities as a resource around which to generate experiences for visitors. On the other hand, tourism activities lead to the transformation of the coastal environment and can reduce fishing resources or access to them, the increase in demand for fishery products can promote their production reaching unsustainable rates, tourist activities can also compete for resources (such as labour) with traditional productive activities, while new populations and urban growth are attracted by new employment opportunities, and, finally, tourism can promote gentrification and touristification processes.

The variability of the relationships between tourism and fishing activities in the environment of massive sun and beach tourist destinations is very wide. But, at the same time, it is common that direct relationships between tourists and fishermen, as guests and hosts, occur only at a very superficial level. Relationships between visitors and the cultures, realities and local population of the destination has been defined as inauthentic. Many are the authors who have marked the relationships between tourists and residents as a parody based on their commercial, accessory, transitory and superficial character [1]–[5].

Although professional fishermen may be constantly surrounded by tourists and tourist services, they seem to be part of the landscape. In this type of massive destinations, local populations tend to depersonalize tourists through their identification with certain stereotypes. Tourists are part of their environment, but they are mostly conceived as homogeneous groups of people with interests and perspectives very different from those of

WIT Transactions on Ecology and the Environment, Vol 256, © 2022 WIT Press
www.witpress.com, ISSN 1743-3541 (on-line)
doi:10.2495/ST220041

the local populations, who will be replaced in a short time by other groups of similar and depersonalized individuals, in a sort of continuous parade. In general, tourists are not identified for locals as subjects with whom to establish genuine personal relationships.

On the other hand, the perception of tourists is conditioned by the experience of an extraordinary space and time. A vacation context marked by the transcendental objective of having fun, contact with exotic environments and cultural practices, as well as the development of recreational activities that allow them to disconnect from their daily experience to return home with new motivations and experiences to remember and share. Tourists consume the destination conditioned by a set of stereotypes that have motivated them to select the environment to visit. The need to confirm these stereotypes has important implications for how they perceive and establish relationships with the destination and its local populations. In general, the tourist will fix his attention on certain exotic elements, hindering the authenticity of the contact and even distrusting the real nature of what contradicts what configured his image of the destination in origin. [6], [7]. The tourist gaze [8] will be directed towards those elements of the destination encouraged by the promotion and interests of the tourist industry. Tourists will focus the lens of their cameras towards these elements, ignoring many other aspects of the reality of the destination visited. In many massive coastal destinations, the fishing activity remains outside these stereotyped constructions, or they constitute complementary elements that contribute a certain degree of exoticism and authenticity to the tourist image of the destination.

Even so, there are many examples in which fishing activities have been directly incorporated into the stereotypical constructions and the list of tourism products of the massive coastal destinations [9]. In these cases, this produces positive effects on the economic exploitation of tourism by local communities, strengthening the possibilities of interaction with tourists and promoting tourist experiences of great value for them. The different forms of inclusion of fishing activities in tourism are presented as strategies for economic diversification responsible with the culture and the territory, based on the enhancement of the historical productive activities of fishing collectives. Trying to cover the various meanings mentioned in the legal framework of the different Spanish territorial areas where it is applied, Moreno Muñoz [10] defines fishing tourism as those "activities developed in a coastal environment, on board or outside fishing vessels, with the aim of diversifying the economy and adding value to a singularity that positively affects the fishing and service sectors in traditional coastal areas through economic consideration, to disseminate the seafaring culture, its traditions, the work in the marine environment and port facilities, the associated material and immaterial heritage and the fishermen's way of life".

Although with different results depending on the locations where fishing tourism has been implemented, we can find successful applications in America, Asia and Europe [11], [12]. The European experience shows that professionals linked to fishing tourism tend to focus on land-based activities, such as gastronomic activities or visits to factories and museums, as opposed to less use, especially in Spanish cases, of fishing vessels [13], [14]. The possibilities of using fishing activities as a tourist experience are very varied. They include sea and land products, active and passive, related to tradition or cutting-edge technology. Given its characteristics, it is understood that fishing tourism requires specific regulations for its development.

In this paper, we analyse the potentialities and conflicts surrounding the possibilities for the development of fishing tourism in the Canary Islands, a destination clearly identified with forms of sun and beach mass tourism.

## 2 FISHING TOURISM AS A SUSTAINABLE OPPORTUNITY

The European Union's Blue Growth strategy specifically promotes proposals for the sustainable development of coastal destinations based on the creation of small and medium-sized marine tourism enterprises (among others), which diversify tourism demand and promote its deseasonalization, with an impact on the reduction of the carbon footprint and the environment. However, on too many occasions these interventions are conceived more as neoliberal resource privatization strategies [15], [16] than as a conservation tool and a means to promote quality of life.

In the Spanish case, with the original initiative of the Galician fishermen's guilds, regulations, projects and specific actions have begun to emerge, trying to break the limitations that prevent the correct development of this type of products. The main blockages perceived by the affected agents seem to be the lack of specific regulation, the inadequate infrastructure and the lack of training and experience of fishermen in touristic sector. The potential positive effects of the development of fishing tourism activities seem clear. On the one hand, the popularization of the activity through heritage and dissemination of the fishing culture and its traditional practices provides an added value compared to the usual sport fishing trip in commercial boats [17], which can promote the prestige of the profession, affecting the fishermen's standard of living and quality of life, and fostering their self-esteem [18], [19]. The consolidation of the product would mean an increase in the usual income and an attraction that would contribute to avoid the transfer of status, capital and personnel, due to the characteristic abandonment of traditional professions of tourist destinations [20] and the aging of the fishing population, favouring generational replacement. Precisely, economic diversification due to the complementarity of this activity, together with its strategic use to fix the fishing population and attract new professionals, appear as the main motivations for promoting this type of enterprise in Spain [14]. In the rest of Europe, on the other hand, its interest is centred on its use as a strategy for taking advantage of and adapting to the new tourist demands derived from the dynamics of touristification of coastal destinations [11].

In addition, this type of tourism activities favours environmental education on coastal environments, both for local populations and visitors [21]. The decrease in fishing trips and times, as well as the reduction of catches during tourist activity, would imply a decline on fishery resources pressure [22], [23]. In fact, the search for strategies to reduce catches is among the main objectives of the European Union's Common Fisheries Policy, promoting interaction between fishing and tourism [9].

Coordination between administrations, scientists and fishing organizations in reconciling the fishing world and tourism in a sustainable way must be an indispensable requirement from the outset. Real participation, beyond consultation, must be accompanied by appropriate technical, regulatory and training support. In this context, it is essential to support horizontal collaboration in the offer of, at least, professional fishermen, accommodation providers and catering professionals with a differentiated gastronomy in accordance with the seafaring culture [11]. At the same time, the complementarity of these activities with other resources linked to the visibility of cultural assets related to the world of fishing (tourists routes, museums, interpretation centres, etc.) will contribute to the objective of economic profitability, socio-cultural equity and heritage sustainability [17]. The lack of collaboration and the shortcomings in the creation of an organized and integrated project that considers the specific characteristics of each locality has revealed structural tensions between artisanal and sport fishing or tourist agencies on many occasions [24]. However, an adequate management of the valorisation of knowledge and practices of fishing communities not only involves the activation of resources that can benefit society [25], but also favours the management of tourism development in coastal destinations [26].

## 3  FISHING TOURISM REGULATION IN THE CANARY CONTEXT

Canary Islands organizations and administrations have been showing interest in the implementation of this activity for decades with little success. Although it has great potential for the development of fishing tourism (wide representation of traditional and artisanal fishing techniques, large tourist influx, good annual weather conditions, etc.), no business venture in this regard has been carried out. Recently, Law 15/2019 of 2 May about Canary Islands Fishing has been enacted, which seems to open the door to the implementation of these products in the islands. However, the activity presents difficulties in taking root as an economic diversification strategy in the archipelago, despite its explicit search for differentiation strategies and mechanisms to overcome its current dependence on tourism.

The reasoning included in the preamble of this law refers to its necessity based on the historical demand of the fishing sector for a regional regulation of complementary economic activities related to fishing and aquaculture. The search for diversification strategies for Canarian fishermen has been justified by the Coastal Action Groups themselves for decades [27]. During the 1950s and 1960s, it was common for professional fishermen to sporadically, on weekends and holidays, take tourists and locals on board to supplement their income. Currently, many fishermen use the rental of houses or the operation of bars or restaurants where their fresh fish is offered to increase their income, although both services lack the potential of the added value that their offer could have linked to fishing trips [27].

In addition to emphasizing the complementary nature of these activities (they may not provide more than 40% of the total income of professionals), the Canary Islands Fishing Law provides that fishing tourism activities must be oriented towards the objectives of disseminating heritage (traditions, trades, gastronomy, culture, arts, gear and techniques), enhancing the social, economic and technological evolution of these communities. The regulations also stipulate the essential requirements for the development of the activity. Among them, the need to have certain licenses, authorizations and permits can be highlighted. But especially relevant are the requirements for a boat to be able to carry out tourist fishing activities (minimum size of the boats, safety conditions, bathroom on board, etc.).

## 4  FISHERMEN'S PERCEPTION OF PESCATURISM:
## POTENTIALITIES AND CONFLICTS

The perception of fishing professionals regarding the possibilities of fishing tourism in the Canary Islands has been approached from a qualitative approach through: (i) observation of dissemination activities of the Canary Islands Fishing Law 15/2019 and the possibilities of business entrepreneurship organized by the Cabildo of La Palma, the Coastal Action Group of La Palma and the Coastal Action Group of Tenerife in 2019; (ii) in-depth interviews with professional fishermen in the locality of Punta del Hidalgo, on the island of Tenerife; and (iii) ethnographic fieldwork during seven months, in which participant observation techniques and in-depth interviews were used in the main fishing localities of Fuerteventura.

Previous studies [28] indicated an enormous predisposition of Canarian professional fishermen towards the implementation of tourist fishing, although with certain misgivings. However, fishermen maintain that the current regulations imply a blockage for the development of the activity. The main argument for this consideration is based on the fact that the regulation does not take into account the specificities of the Canarian artisanal fleet. The small size of most of the traditional inshore fishing boats of the Canary Islands can be an added value to the product, both for the picturesque appearance of the boats, as well as for the ease and safety implied by the short distance from the coast in which they usually fish. However, their size is one of the main limitations for the development of fishing tourism in the region. This is because it makes it impossible to comply with the regulations designed for larger vessels: minimum crew requirements, safety structures such as handrails and lifeguards, toilets on board, etc.

The perception of Canarian fishermen of the laxity of these regulations in other international contexts causes a feeling of helplessness and incomprehension in these agents, who try in vain to refute the administrations with ad hoc arguments against a regulation considered generalist and ambiguous, which does not take into consideration local particularities.

> "The toilet... I had a meeting and it is complex. Because it has a measure and I was not in agreement with that. I have filed a complaint about that. If it is done, I am not in agreement at all, because then the whole northern fleet of Fuerteventura is left without fishing tourism. The Azorean boats are like mine, smaller boats and they are going with fishing tourism. The women's association of the Azores is taking pesca turismo, they are the ones that started first. And they are 6-m boats. What are you telling me?! Because one day I was with the merchant marine in Tenerife, and they told me that no, I should have a cabin at the front and put a tiny bathroom like the one in the caravans inside, with its little door (...) Those who are going there [initiatives similar to pesca turismo developed with commercial pleasure boats], half of them have nothing and you see them coming and going in a zodiac. And I at least have a chemical toilet inside the cabin. And they said it was a danger..."
>
> (Senior Patron of the fishermen's guild 1)

> "Is that a problem? What can I tell you... on a trawler. As long as it doesn't go too far... Because if you put it, as is being done in France and in those places... France is already working with tourist fishing; Italy is working on a project and the Azores are working on tourist fishing. And the guys said 'no, we put a glass window and no one will pass from here to there', they built it and you can't pass. One of our sardine seiners, well, look, it's small, you get inside the cabin and: 'Hey, you can't go out until the fishing is done!' So... And with me, on my boat, you give him a rod, two people... I'm not going to take 10, I can take one or two at most. For me, the one who pays 50€ solves me half a day."
>
> (Professional fisherman 1)

(All interview verbatims originally in Spanish have been translated literally.)

The Canary Islands law expressly prohibits the development of fishing activities by tourists in professional boats. This prohibition does not extend to other types of tourist activities not related to professional fishing and contrasts with the numerous companies that offer sport fishing activities in recreational boats with certain safety and habitability requirements. This issue reinforces the scepticism of the professionals regarding the way in which decisions are made, increasing their feeling of disregard.

> "The most absurd of all this is that you get the title of Pleasure Craft Skipper or any title of yacht skipper or yacht captain or whatever you want, and you get it today and tomorrow you can go sailing. And I can have been going to sea for 20 years as a sailor... whatever you want, and I get my local skipper's certificate and I have to do 6 months on the bridge, 6 months on deck and 6 months on engines. With all the experience I have! I have the sailing license

that says that I have been sailing for 5 years, 10 years or 20 years, [and they tell me:] 'No, what you have to do is this'. They don't give it to you. But tomorrow you get the title of Pleasure Craft Skipper... or whatever you want, tomorrow you buy the yacht... and you can go around the world if you want. But I don't understand that..."

(Professional fisherman 4)

The fishermen concerned consider that the fishing tourism product could be much more attractive if tourists were allowed to participate in traditional fishing practices, understanding that a more flexible regulation, similar to the Italian one, would facilitate the growth of the activity and its associated socio-environmental benefits. In this way, it is argued that the participation of tourists in fishing would not only constitute an added value, but would also favour economic diversification, contributing to the self-esteem of fishermen and to the reduction of pressure on fisheries. The adequacy of such arguments to those of public administrations, point to the strategic appropriation of institutional discourses on socioeconomic and environmental sustainability, redirecting them in their favour. This practice has been observed in other contexts of territorialization conflicts in local Canary Islands populations as a tool to legitimize their postures and position themselves in the decision-making arena in the management of their territory [29]. A discourse that sometimes even includes the suitability of their justifications for the general tourist offer.

"Here, when tourism started, what we had were fishing boats. People came to the beaches and went out with the people from the villages, they went and learned... And that is what was asked for. I mean, I am a sailor and I can take two people today and two people represents a profit of 100 euros? Well, that will mean that I will not need to fish so much and the little fish will rest. It is compatible with something! (...) And, for example, if you are a tourist, 'You are going to go to sea, do you want to go with a professional sailor and spend a day...', 'Yes'. It's not the same to say, 'I'm going on a boat and I'm going with somebody'. You get that anywhere. But a fishing boat, you tell them how to fish, how to... And it's different even in the sense of... even the people, you meet the people afterwards and you say 'Well...' and you make bonds with that family, with those people, they recommend them to someone else, they come and you invite them to eat at your house, you go to that country and they invite you, they introduce you to the other one and send you to the other one... and so we all create bonds of friendship. (...) These are the best ambassadors we have in relation to tourism. They are not the tour operators, nor are they anything else: they are those people. You take 10 and 10 inform 10,000. And the other one informs so and so. And those people already come recommended by those people, not by tour operators or anything else. And they come straight to it and leave money directly here (...) The nice thing is that, that you have contact with them and say 'Look, I'm not going to see it as making money, but as a way of living and opening myself to the world'. It is a way of life that is not with the ambition of saying 'I am going to get rich and buy a cruise ship'. No, it's living and going back days and years, and meeting people, and enriching yourself in culture, getting to know others, talking with some, with others..."

(Senior Patron of the fishermen's guild 2)

"It would be fine with me because the fishing is... decreasing. (...) And you have to be fishing from sunrise to sunset. It was not like before, when we used to get there and do nothing more than fishing, a good catch, and at two o'clock in the afternoon we would go ashore with a full catch, with the fish caught. It is noticeable that fishing is decreasing (...) We have to consider, for example, to take tourism on all our boats and leave a year without fishing. That is, you go with them and throw a fine nylon so that they... you know? But we don't go for all I can, but to spend a little while with them. Or to see the dolphins, to see the whales... and all that. And it seemed to me that [if we do that] on all ships... and leave the sea for a year. We... look, every day, with four or five tourists it solves us. They are amazed if you take them on a boat... Damn! How many have come asking about the boats here? The tourists. They don't want to go to where the tourists are, they want to go on a fishing boat. A lot of them!"

(Professional fisherman 8)

Even those fishers who are not interested in participating in fishing tourism understand the potential of the activity. Many fishermen allude to past times when professional fishermen took tourists aboard their boats to fish and are aware of its usefulness in supplementing the fishermen's income with less productive strategies. In contrast, those professionals who are more productive (with longer working days, at greater depths, etc.), consider that taking tourists on board may be an inconvenience that hinders their work.

"The one who lives by fishing doesn't really need to take two people on board who are of no useful for you. And then, more than anything else, the stomach in fishing when they are at sea. We sometimes... the smaller boats had Germans on board. A lot of years ago they always took them to fish. And that was one thing... And many times they were around here and so on... But we can't take them. Many people prefer to go fishing in a professional boat rather than in a sport boat... But I, for example, would not be interested, at least it's not for me. People who fish less or more sporadically might be interested. We are talking about conditions that are sometimes... hard. If you are going to put a tourist there at... with 25 knots of wind... I must take him back to land and I lose a day of fishing. It is not worthwhile."

(Professional fisherman 7)

The safety discourse is another recurrent argument perceived by artisanal fishermen as an ambiguous impediment imposed by the institutions. Once again, we can see how the discourse of danger is appropriated by these agents when it comes to positioning themselves as a group worthy of participating in the activity, deserving of a leading role granted by the knowledge of the profession and the tradition and uniqueness of their techniques.

"Danger, in a boat of mine, to go fishing for 'vieja' [*Sparisoma cretense*, fish with high local significance]? That 'vieja' is fished with a meter of depth in the shore of the island... On one hand, that I don't go out with bad weather; and in the other hand, you are always there, at two meters. The 'vieja' is fished in the shore because you can't fish it where there are waves. And we fish 'vieja' at one meter deep, one and a half meters deep and 20 meters from the shore. Is that danger to take one or two guys, and with me three? Man, I'm not going to take 10 guys on a 6 meters boat! But two people with you?"

(Professional fisherman 1)

Some professional fishermen interviewed were sceptical about the potential of this type of tourism product, understanding that if visitors cannot participate in the fishing operations, the activity resembles a boat trip, a practice with wide competition, marketed by specialized entities and at very low prices. Other professionals, on the other hand, show the discourse of fishing specificity, both when fishing and for their knowledge of the environment, to legitimize their position in favour of the compatibility of their profession with tourism. Both may be key aspects to their empowerment and the enhancement of their daily cultural practices.

> "That area, above all, I love that area. I go fishing and I usually stay on my boat... I usually go that way during the summer. And I go fishing and I'm in that area one day, I jump ashore, I arrive in the afternoon, and I have my little meal there on the sand... but fish or anything else. I jump into the water... it's a small cove with a white sandy bottom (...) And I'm there in the rocky area. I don't go fishing the next day, but I say to myself 'come out the next day, there's a paradise here'. And then you have some puddles, some lakes in between... You take tourists, foreigners, you put them there and they go crazy"
>
> (Senior Patron of the fishermen's guild 2)

Regarding the link with other types of tourism, only the possibility of developing diving activities related to aquaculture tourism, apart from the complementary nature of coastal walks or gastronomic activities, was mentioned. This contrasts with the potential for linking artisanal fishing with other forms of tourism developed in the coastal or marine environment, apparently very broad. The fact that this is not evident may be related to one of the key limitations perceived by some fishermen: their low involvement and knowledge of the tourism sector. The information gathered suggests that their interaction with tourist activity is very limited. Although professional fishermen in the Canary Islands are constantly surrounded by tourists and tourist-oriented services, they seem to be part of another layer of reality with which they interact only superficially, as just another component of the landscape.

The low knowledge of the tourism sector is related to limitations in designing attractive products adapted to different visitor profiles, interaction with tourists in other languages, and communication and marketing strategies for their products. Professionals in the fishing sector are hesitant about the way in which their offers could reach the demand, as well as the management of this activity. Giving a relevant role to fishermen and the presence of cooperatives can be an effective tool when proposing and advertising offers, coordinating activities according to the capacity and fishing strategies of the vessels, as well as for administrative, financial or commercial tasks. In this sense, to increase the chances of success, the activity should begin by being implemented in those localities where there are already cooperatives or operational fishermen's associations that can offer the necessary support.

The mistrust of Canarian fishermen due to the feeling of lack of real involvement in decision making is related to their scarce entrepreneurial practice and to past experiences in which the benefits of exogenous implementations have fallen in few hands or directly outside the community [29]. Their fears thus lie in the possibility of a poorly recognized effort that would mean, in the medium term, opening the door to unintended harms. This concern has been observed in other contexts [30] where insufficient or no local participation in development projects, rigid bureaucracy and administrative and institutional incoordination hinder the success of new initiatives, while hindering the dynamics of adaptation and appropriation by local populations [19], [31]. The feeling of imposition and lack of transparency, the impersonalization implied by the generalization of regulations and the wariness about the modification of certain historical patterns that fishing tourism entails [19], [28], puts on alert a collective that distrusts an administration that unilaterally regulates its daily life and fears the effects of the loss of control over the activity.

## 5 CONCLUSIONS

Fishing tourism has the potential to complement and diversify the productive activity of this sector. On the one hand, it can be a tool to improve the quality of life and self-esteem of the population linked to fishing by enhancing the value of their practices. On the other hand, it can be a key factor in reducing pressure on resources and promoting tourism that respects the environment and local culture. For this, it is essential to have an adequate management that takes into account the demands and perceptions of the main agents involved, taking advantage of the inertia of fishermen's collective organizations that can offer their support and exploring ways to help overcome certain limitations contemplated by professionals in the sector. The research shows an active position of fishing professionals, in which institutional discourses on socioeconomic, environmental and cultural sustainability are apprehended and used strategically by the fishermen's collective to justify the appropriateness of their intervention in the decisions at stake.

Appropriate regulation is a basic requirement for the development of an activity that links artisanal fishing with tourism, but it must be complemented by active training and incentive measures to ensure that this set of tourism proposals effectively results in the sustainability of the areas concerned. An imposing approach, which does not contemplate an effective participation of fishing populations in decision-making and does not reinforce the processes of appropriation of symbolic identification referents by fishing professionals, can have the opposite effect, accelerating the performativity of traditional practices, the trivialization of identity assets and the undervaluation of the activity, as well as opening the door to a new possibility of territorial consumption, not necessarily complementary.

## REFERENCES

[1]    Getz, D., Impacts of tourism on residents' leisure: Concepts, and a longitudinal case study of Spey Valley, Scotland. *The Journal of Tourism Studies,* **4**(2), 1993.

[2]    De Kadt, E., *Tourism: Passport to development?* Oxford University Press: New York, 1979.

[3]    Crick, M., Representations of international tourism in the social sciences. Sun, sex, sights, savings and servility. *The Sociology of Tourism: Theoretical and Empirical Investigations*, eds Y. Apostolopoulos, S. Yiannakis, S. Leivadi & A. Yiannakis, Routledge: London, 1996.

[4]    Santana, A., *Antropologia do turismo. Analogias, encontros e relaçoes*, Aleph Publicações: Rio de Janeiro, 2009.

[5]    Barretto, M., *Turismo y cultura. Relaciones, contradicciones y expectativas*, Pasos Edita: Canary Islands, 2007.

[6]    Faulkner, B., Moscardo, G. & Laws, E., *Tourism in the 21st Century: Lessons from Experience*, Continuum: London, 2000.

[7]    MacCannell, D., Remarks on the commoditification of cultures. *Host and Guest Revisited: Tourism Issues of the 21st Century*, eds V. Smith & M. Brent, Cognizant Communication Corporation: New York, 2001.

[8]    Urry, J., The tourist gaze 'revisited'. *American Behavioral Scientist*, **36**(2), pp. 172–186, 1992.

[9] Lois González, R.C. & De los Ángeles Piñeiro, A., Fishing tourism as an opportunity for sustainable rural development. The case of Galicia, Spain. *Land*, **9**(11), 2020.

[10] Moreno Muñoz, D., Turismo marinero ¿Complemento del modelo turístico de sol y playa en la Región de Murcia? *Actas del X Congreso Internacional de Turismo Rural y Desarrollo Sostenible*, eds X.M. Santos, P. Taboada & L. López, Universidad de Santiago de Compostela: Santiago de Compostela, pp. 411–420, 2016.

[11] Pardellas, X.X. & Padín, C., La nueva demanda combinada de turismo litoral y turismo pesquero: Motivaciones y efectos. *Cuadernos de Turismo, 32*, pp. 243–258, 2013.

[12] Moreno Muñoz, D., Aportación a los conceptos de turismo marinero/pesquero y pesca-turismo. *Cuadernos de Turismo, 42*, pp. 385–396, 2018.

[13] Pardellas, X., Padín, C. & Aboy, S., Turismo pesquero: Experiencias en Europa y España. *Papeles de Economía Española, 128*, pp. 221–227, 2011.

[14] Centro Tecnológico Del Mar (CETMAR), *La Pesca de Bajura: Situación en 2007 y Perspectivas para una Orientación Sostenible,* CETMAR: Pontevedra, 2010.

[15] Harvey, D., Crises, geographic disruptions, and the uneven development of political responses. *Econ. Geogr.*, **87**(1), pp. 1–22, 2011.

[16] Cortés, J.A. & Apostolopoulou, A., Against neoliberal natures: Environmental movements, radical practice and the right to nature. *Geoforum*, **98**, pp. 202–205, 2019.

[17] Pascual, J., Las investigaciones sobre la pesca en Canarias: entre las reservas marinas y las nuevas formas de pescaturismo. *Pasos. Revista de Turismo y Patrimonio Cultural, 2*(2), pp. 295–306, 2004.

[18] Varela, M. (ed.), *Unha Estratexia Marítima para Galicia.* Galaxia: Vigo, 2010.

[19] Herrera-Racionero, P., Miret-Pastor, L. & Lizcano, E., Viajar con la tradición: los pescadores artesanales ante la pesca-turismo en la Comunidad Valenciana (España). *Cuadernos de Turismo, 41*, pp. 279–293, 2018.

[20] Santana Talavera, A., Turismo, empleo y dependencia económica: Las estrategias de las unidades domésticas en dos poblaciones pesqueras (Gran Canaria). *Eres (Serie de Antropología), 2*(1), pp. 25–38, 1990.

[21] Miret Pastor, L., Muñoz Zamora, C., Herrera Racionero, P. & Martínez Novo, R., Análisis regional del turismo pesquero en España. *Revista Turismo em Análise, 20*, pp. 23–28, 2015.

[22] Nicolosi, A., Sapone, N., Cortese, L. & Marcianò, C., Fisheries-related tourism in southern Tyrrhenian coastline. *Procedia Social and Behavioral Sciences*, **223**, pp. 416–421, 2016.

[23] Croes, R., Ridderstaat, J. & Van Niekerk, M., Connecting quality of life, tourism specialization, and economic growth in small island destinations: The case of Malta. *Tourism Management*, **65**, pp. 212–223, 2018.

[24] Chen, C. & Chang, Y., A transition beyond traditional fisheries: Taiwan's experience with developing fishing tourism. *Marine Policy*, **79**, pp. 84–91, 2017.

[25] Brookfield, K., Gray, T. & Hatchard, J., The concept of fisheries-dependent communities. A comparative analysis of four UK case studies: Shetland, Peterhead, North Shields and Lowestoft. *Fisheries Research*, **72**, pp. 55–69, 2005

[26] Rubio-Cisneros, N.T. et al., Poor fisheries data, many fishers, and increasing tourism development: Interdisciplinary views on past and current small-scale fisheries exploitation on Holbox Island. *Marine Policy, 100*, pp. 8–20, 2019.

[27] Grupo de Acción Costera Fuerteventura (GAC), *Estrategia de desarrollo local participativo*, UE-Gobierno de Canarias, 2021.

[28] IEO, *Caracterización del Banco de La Concepción. Informe del Instituto Español de Oceanografía-Centro Oceanográfico de Canarias, LIFE+ INDEMARES.* Fundación Biodiversidad: Madrid, 2013.

[29] Díaz, P., Cambio cultural e imagen turística. Efectos de los procesos de reconfiguración y consumo de imágenes proyectadas. Doctoral thesis, Univ. Pablo de Olavide, Seville, 2015.

[30] Herrera-Racionero, P., Lizcano, E. & Miret-Pastor, L., 'Us' and 'them': Fishermen from Gandía and the loss of institutional legitimacy. *Marine Policy*, **54**, pp. 130–136, 2015.

[31] European Commission, Rebuilding cod and hake stocks. *Fishing in Europe*, **18**, 2003.

# RURAL TOURISM UNDER THE NEW NORMAL: NEW POTENTIALS FROM A JAPANESE PERSPECTIVE

YASUO OHE

Tokyo University of Agriculture, Japan

## ABSTRACT

This paper explores the new directions of rural tourism and potential demand characteristics under the new normal resulting from COVID-19 based on an internet questionnaire survey targeted at inhabitants of large cities in Japan. Specifically, this paper reveals the features of potential demand targets, required rural attractions, and essential facilities for micro-tourism, defined herein as tourism around the neighbouring areas of tourists and workcation as the combination of remote work and vacation. Findings were that micro-tourism attracts interest particularly among the mature and workcation attracts interest among younger generations. This is in contrast with the fact that the young have shown lower interest in rural attractions and life than older generations. The natural environment, rural heritage, and local food are three common major attractions for the two types of tourism. Respondents expressed that basic business facilities and restaurants should be available for a workcation. COVID-19 promoted digitalization such as online travel agents and extended the area of e-hospitality based on digital technology in addition to on-site traditional face-to-face hospitality. Thus, operators should understand that hospitality starts at the device terminal and explore how to integrate these two types of hospitality well.

*Keywords: COVID-19, rural tourism, micro-tourism, workcation, e-hospitality, resilience.*

## 1 INTRODUCTION

COVID-19 plunged the globe into a huge socio-economic tailspin and particularly inflicted an unprecedented negative impact on the tourism and hospitality industries. The successful inbound promotion policy in Japan was choked off due to border closures. Domestically, there are no clear prospects for the recovery track of tourism and hospitality businesses due to repeated declarations of states of emergency as of January 2022. The tourism and hospitality industries are vulnerable to wars and conflicts, political turmoil, and natural disasters. This pandemic has joined the list of threats.

The emerging variants and waves of surging infections, one after another, has inevitably shed light on the issue of resilience. It is true that new buds of activities are emerging in many arenas. From this perspective, many studies have evaluated the impact of COVID-19 and the recovery track of tourism and hospitality from qualitative and quantitative approaches, examples of which are Fotiadis et al. [1], Sharma et al. [2], Škare et al. [3], and Qiu et al. [4]. With respect to rural tourism, Ohe [5] conducted a short review of the rural tourism situation in Japan, and Polukhina et al. [6] reviewed a Russian case. Ohe [5] considered new possibilities for rural tourism as a result of COVID-19. That, however, was a preliminary report and further investigation of new demands related to tourism in the countryside is needed.

Thus, this paper explores potential emerging demands for rural tourism under the new normal and future evolution of rural tourism based on responses to a questionnaire survey of metropolitan inhabitants after COVID-19. Ohe defines rural tourism from a microeconomic viewpoint, which mentioned that rural tourism is activities that internalize positive externalities generated by multifunctionality of agriculture [7], [8]. Specifically, this paper sheds light on two potential demands for rural tourism. The first is micro-tourism, which is defined as tourism mainly within day-trip distance by the author. The second is the

WIT Transactions on Ecology and the Environment, Vol 256, © 2022 WIT Press
www.witpress.com, ISSN 1743-3541 (on-line)
doi:10.2495/ST220051

"workcation", which means the simultaneous engagement in the combination of telework and vacationing during a countryside stay by the author. These two types have attracted growing interest, which leads to prospects for the future evolution of rural tourism after COVID-19 [9]. Finally, policy implications will be suggested.

## 2  IMPACTS OF COVID-19 ON GLOBAL AND JAPANESE TOURISM

### 2.1  Present situation of rural tourism programs in Japan and the COVID-19 impact

First, on the COVID-19's impact on inbound tourism to Japan, tourist arrivals had surged from late 2010 to 2019 but subsequently plummeted from 31.88 million arrivals in 2019 to 4.12 million in 2020 and devastatingly to only 0.23 million up to November 2021 due to COVID-19 according to JNTO as of January 2022.

Beginning in 2017, rural tourism in Japan entered a new phase with the inauguration of a novel program called "Nohaku", i.e., farm/rural stay [8].

What differs among other things from the former type of rural tourism program, i.e., green tourism, is that the novel program is clearly oriented toward viable economic activity although its community-based nature has not changed [10]. Before the advent of the new program, rural tourism heavily depended on school trips for which quality of service did not matter much and which was not targeted at tourists such as the free independent traveller (FIT) and up-market inbound tourists from abroad [11]. For this reason, the former activity resulted in low-average spending per guest and income not sufficient to sustain a young-generation family as a full-time occupation despite how busy they could be with a group of students [10], [11]. Therefore, rural tourism has become directed toward increasing viability as a farm business.

From this perspective, the new program intends to raise the value of the service, to make it a viable business, and to target not only more domestic FITs but also inbound tourists from abroad. In this context, this program is consistent with the national inbound-tourism policy framework. The number of Nohaku areas approved by the Ministry of Agriculture, Forestry and Fisheries (MAFF) was 554 as of the end of 2021 [9].

Nevertheless, COVID-19 inflicted serious damage to the Nohaku business. From 2017 to 2019, domestic arrivals at Nohaku sites increased from 4.748 million to 5.515 million and inbound arrivals increased from 0.286 million to 0.376 million. In total, there were 5.892 million, or close to 6 million, arrivals in 2019. The average number of arrivals per area was about 10 thousand, with domestic tourists accounting for 93% to 94%. This means that the inbound market for rural tourism has not been realized although the new program promoted inbound tourism to rural areas. This high dependency on the domestic market works as a safety net under the situation of border closures because of the pandemic. Now let us look at this aspect specifically. In 2020, after the pandemic began, the number of arrivals dropped to 3.886 million domestic tourists and 0.019 million tourists from abroad. The inbound number markedly plummeted by 95% while there was a one-third drop in the domestic market compared with 2019. Thus, taking the small percentage of inbound tourist in consideration, in the end there was a 30% drop in total stays. Although that 30% reduction cannot be considered light damage, it is also true that the domestic market prevented that business from devastation.

The MAFF report said that the number of meal services largely increased from an average of 18.3 menu items in 2019 to 23.7 menu items in 2020 [11]. This is probably because operators adjusted the food services aiming at day-trippers to compensate for the decrease in demand from other sources. To put it differently, operators expected increased opportunities related to micro-tourism and restaurant activity.

Further, regarding utilities, the utilization of non-contact booking sites, that is online travel agents (OTAs), has significantly grown from 30.7% in 2017 to 63.9% in 2020 (Table 1). Introduction of Wi-Fi slightly increased during the same period.

Table 1:  Provision of utilities during "Nohaku". *(Source: Report on Nohaku, October 2021, MAFF.)*

| Item | 2017 | | | 2020 | | | 2020/2019 ratio |
|---|---|---|---|---|---|---|---|
| | Yes | No | % Taken-up | Yes | No | % Taken-up | |
| Using OTA | 63 | 142 | 30.7 | 354 | 200 | 63.9 | 2.08 |
| Wi-Fi availability | 128 | 77 | 62.4 | 367 | 187 | 66.2 | 1.06 |
| Western style toilet | 160 | 45 | 78.0 | 401 | 153 | 72.4 | 0.93 |
| Foreign language | 108 | 97 | 52.7 | 281 | 273 | 50.7 | 0.96 |
| Number of approved areas | 205 | | – | 554 | | – | 2.70 |

Note: Years are fiscal years from April to March.

To summarize, under the pandemic, Nohaku operators increased meal service menus and utilization of OTAs, which are necessary for micro-tourism and the workcations mentioned below.

## 3 IDENTIFYING NEW DEMANDS FOR RURAL TOURISM UNDER THE NEW NORMAL

Under the pandemic, those involved in various aspects of rural tourism recognized that new rural orientations were emerging among urban dwellers [5], [9]. Thus, the author investigates these orientations as potential demand by focusing on micro-tourism and workcations based on the results of a MAFF-supported questionnaire survey to inhabitants of large cities in Japan. First, we shall clarify the definitions of these two types of tourism, which have not yet been described by MAFF. Based on the definition above, micro-tourism is considered as domestic tourism behaviour manifested by visiting places within areas in which daily life is carried out, i.e., daily heritage sites. People are expected to rediscover their own community resources and values. In this context, rural tourism has many suitable elements for this new kind of tourism activities. For instance, although the natural environment and cultural heritage have been traditionally pointed out as rural tourism resources, they are not always well acknowledged and utilized [9]. Micro-tourism can create opportunities for inhabitants to open their eyes toward their own community. Local resources have value in daily life. Inevitably, micro-tourism is naturally comprised of day or short trips.

Second, the workcation is a synthesis of work and vacation, which reflects the rapid increase in telework or remote work due to COVID-19. Telework or remote work represents a style of work rather than a type of tourism. In this context, telework is a component of workcation. Azuma touched upon workcation as a new phenomenon under COVID-19 in Japan [12]. This presents another tailwind for rural tourism. Working in a spacious facility with relaxing surroundings that presents a lower risk of infection can be an attractive feature of a new lifestyle for urban workers used to a daily journey on heavily packed commuting trains. Thus, the workcation presents new values of rural areas in terms of working in a relaxing and safe environment away from infection [5].

MAFF supported surveys to see if there are potential demands for the two types of tourism. The author uses results of one survey conducted to investigate who has the potential demand, what tourists want to do in the countryside, and issues such as necessary facilities and services. Specifically, data were from an internet questionnaire survey conducted by "Hyakusen-Renma", which operates an online booking site specialized in special interest tourism (SIT) such as temple stays, castle stays, farm/rural stays, etc. This company is an OTA of SIT. This survey targeting a thousand residents in large cities, i.e., Tokyo, Osaka, and Nagoya, from the age of 20 to those in their 70s was conducted in June 2020. Although the return rate was not disclosed, and we should carefully consider this point, the author feels that there is no serious problem in obtaining an overview of the potential of these new types of tourism and related issues from the data provided.

First, Fig. 1 shows that generation wise willingness to make a trip to the countryside. The figure clearly indicates that the willingness was higher in young than older generations with over 70% of respondents in their 20s and 30s expressing a greater willingness in contrast to the elder generations. This point is interesting as it is quite the opposite of what has been assumed, which is that the young are less eager for the rural experience than their elders [5]. To put it differently, the new normal can generate new opportunities and evolution for the rural tourism business.

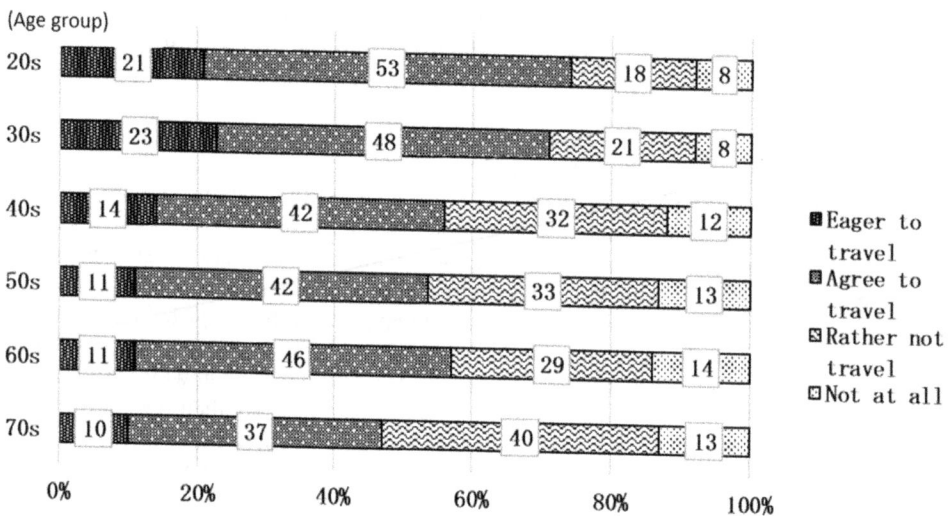

(Age group)

| Age group | Eager to travel | Agree to travel | Rather not travel | Not at all |
|---|---|---|---|---|
| 20s | 21 | 53 | 18 | 8 |
| 30s | 23 | 48 | 21 | 8 |
| 40s | 14 | 42 | 32 | 12 |
| 50s | 11 | 42 | 33 | 13 |
| 60s | 11 | 46 | 29 | 14 |
| 70s | 10 | 37 | 40 | 13 |

Figure 1:  Generation-wise willingness to travel to rural areas. *(Source: Internet questionnaire survey to three Metropolitan inhabitants conducted by Hyakusen-Renma in June 2020.)*

As to the purpose of the rural trip under the new normal (Fig. 2 (multiple choice)), "Rediscovery of neighbouring destinations" accounted for the highest response (62.4%) followed by "as telework site" (30.2%). These two responses demonstrate the potential demands for micro-tourism and workcation, respectively. Of interest is that these orientations are expressed under the new normal. Conversely, we can say that COVID-19 unveiled a potential demand for rural tourism that has not been observed before [8, pp. 38–39]. Thus, we need to further look at generation-wise differences in the selected purposes of the tourism.

The proportion of respondents selecting a workcation was higher in younger generations than in the elder generations: 46% in their 20s, 35% in their 30s, and 25% in their 40s (Table 2). This result demonstrates that a workcation can be accepted among the younger generations. In contrast, micro-tourism does not show distinctive generational differences. Two-thirds of those in their 50s preferred micro-tourism, which means that mature individuals have a great interest in this type of rural tourism. Rural experiences for children, which have been promoted by policy, had only a 10% level of interest except for 28% of respondents in their 30s who would be supposed to have small children.

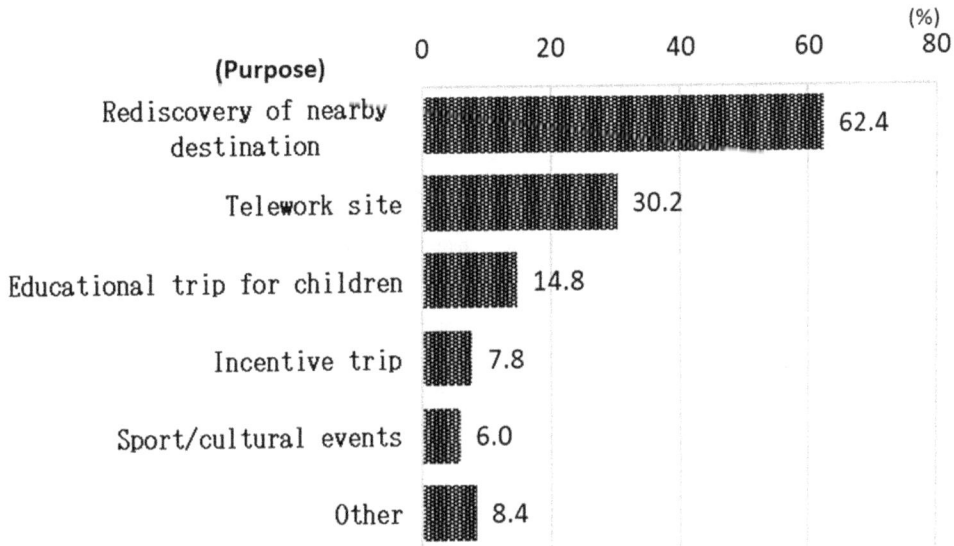

Figure 2:    Purpose of trip to rural areas under the new normal (multiple choice). *(Source: Internet questionnaire survey to three Metropolitan inhabitants conducted by Hyakusen-Renma in June 2020.)*

Table 2:    Generation-wise purpose of rural visit (multiple choice). *(Source: Internet questionnaire survey to three Metropolitan inhabitants conducted by Hyakusen-Renma in June 2020.)*

| Age group | Telework site | Workcation site | Micro-tourism | Incentive trip | Educational trip for children | Sport/cultural events | Other |
|---|---|---|---|---|---|---|---|
| 20s | 15 | 46 | 59 | 11 | 16 | 7 | 5 |
| 30s | 17 | 35 | 61 | 13 | 28 | 5 | 3 |
| 40s | 13 | 25 | 51 | 7 | 16 | 5 | 10 |
| 50s | 13 | 18 | 67 | 5 | 8 | 5 | 8 |
| 60s | 5 | 11 | 66 | 7 | 9 | 6 | 13 |
| 70s | 8 | 7 | 70 | 4 | 11 | 8 | 12 |
| Total | 12 | 24 | 62 | 8 | 15 | 6 | 8 |

To summarize, we can understand that these new types of rural tourism attract different generations: workcation for the young and micro-tourism for the older generations. Further, micro-tourism attracts more than half of those in their 20s and a wide range of generations. Thus, micro-tourism will be widely accepted.

To specify potential demand further, now turning to workcation by occupation. There is a large difference from one occupation to another (Table 3). Among the over the 30% selecting workcation were students (50%), self-employed (35%), full-time company employees (34%), and company executive/presidents (32%). Conversely, interest was much lower by pensioners (2%) and full-time housewives/husbands (11%) due to the lower necessity to telework.

Table 3: Occupation-wise purpose of rural visit under the new normal. *(Source: Internet questionnaire survey to three Metropolitan inhabitants conducted by Hyakusen-Renma in June 2020.)*

| Item | Telework site | Workcation site | Micro-tourism | Incentive trip | Educational trip for children | Sport/cultural events | Other |
|---|---|---|---|---|---|---|---|
| Company employee (full-time) | 15 | 34 | 58 | 16 | 10 | 5 | 4 |
| Company employee (not f-time) | 8 | 21 | 58 | 10 | 8 | 4 | 9 |
| Self-employed | 13 | 35 | 59 | 12 | 8 | 6 | 13 |
| Company owner/manager | 9 | 32 | 50 | 5 | 27 | 23 | 9 |
| Part-time worker | 8 | 16 | 72 | 21 | 10 | 8 | 9 |
| Full-time housewife/husband | 9 | 11 | 68 | 23 | 4 | 4 | 16 |
| Public employee | 7 | 16 | 65 | 19 | 9 | 2 | 5 |
| Pensioner | 2 | 2 | 80 | 10 | 1 | 3 | 8 |
| Student | 20 | 50 | 60 | 0 | 0 | 10 | 10 |
| No occupation | 13 | 11 | 58 | 9 | 2 | 9 | 16 |
| Other | 18 | 18 | 64 | 18 | 9 | 18 | 9 |
| Total | 12 | 24 | 62 | 15 | 8 | 6 | 8 |

Thus, it is safe to say that these new types of tourism attract various segments while students are common in selecting both types of tourism probably because they are time-flexible and more sensitive to work–life balance.

To further explore the preference for workcations, Fig. 3 shows interest toward telework and workcation by the type of jobs. There is great interest in teleworking across job types (Fig. 3). However, the degree of interest is different from one type of job to another. Thus, telework and workcation go well together for some types of work but not for other types. This indicates that there is a difference in demand between telework and workcation, which suggests that the two do not represent an identical market. Specifically, among creators/editors and IT workers/engineers over 30% responded for workcation and over 20% for telework, percentages that are roughly in parallel.

In contrast, nearly half of civil engineering/construction/farm/forest/fishery workers expressed interest in workcation but only 29% for telework. Of teachers/lecturers 42% selected workcation and 17% selected telework, while 38% and 10% of service/sales workers and 36% and 9% medical/nursery/welfare workers selected workcation and telework, respectively (Fig. 3). These results suggest that those with jobs that are difficult to take telework with them and that are stressful expressed a higher interest in workcation.

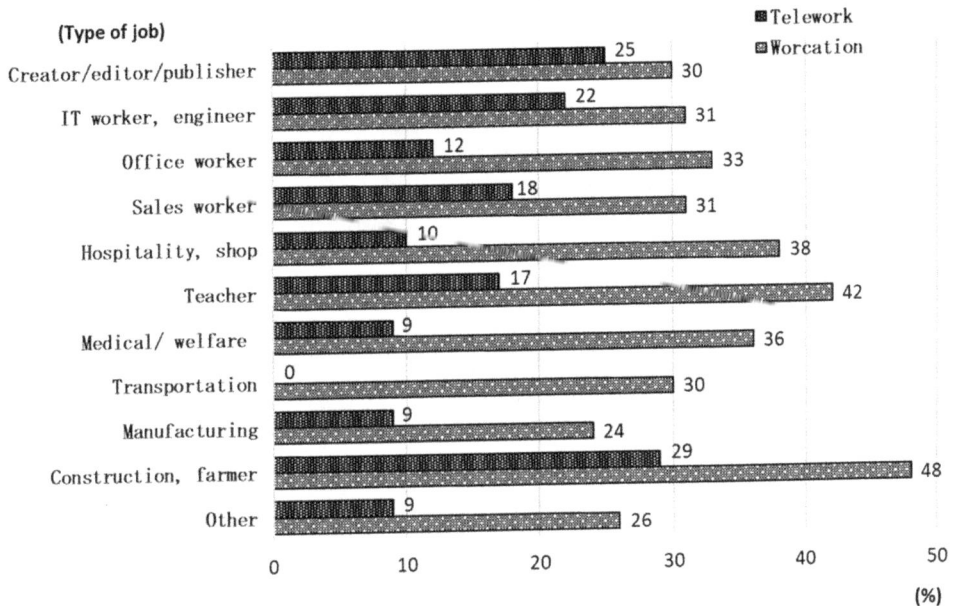

**(Type of job)**

■ Telework
▨ Worcation

| | Telework | Worcation |
|---|---|---|
| Creator/editor/publisher | 25 | 30 |
| IT worker, engineer | 22 | 31 |
| Office worker | 12 | 33 |
| Sales worker | 18 | 31 |
| Hospitality, shop | 10 | 38 |
| Teacher | 17 | 42 |
| Medical/ welfare | 9 | 36 |
| Transportation | 0 | 30 |
| Manufacturing | 9 | 24 |
| Construction, farmer | 29 | 48 |
| Other | 9 | 26 |

(%)

Figure 3:    Purpose of rural visit under the new normal (type of job). *(Source: Internet questionnaire survey to three Metropolitan inhabitants conducted by Hyakusen-Renma in June 2020.)*

To summarize, we can say that workcation provides an opportunity to relieve stress mainly for those in the service industry, which is of social significance in that rural tourism can provide workspace and holiday space together.

Finally, we look at necessary attractions and facilities for a workcation. Fig. 4 illustrates what additional values respondents expected for a workcation (multiple choice). Most popular were rich nature and landscape views and local wining and dining (both 86%), followed by a traditional town, ability to walk to historical sites, and cycling (67%). Thus, it is safe to say that nature, heritage, and food are the three major attractions for metropolitan inhabitants.

Fig. 5 shows the required facilities for a workcation (multiple choice). Basic office facilities such as high-speed Wi-Fi, teleconference gadgets, and photocopy machines must be provided. Following were neighbouring supermarket/convenience stores (67%) and neighbouring restaurants/delivery services (65%). In short, respondents want basic office equipment and nearby businesses that can provide daily necessities. Further, additional utilities were selected such as onsen (53%), i.e., hot springs, dining kitchen (47%), and private space for a family (47%). These utilities will make the stay comfortable and memorable, which may lead to repeat visits in the future.

To conclude this section, it was revealed that necessary conditions for a workcation are basic office equipment and private space for the family in terms of hardware, shops available for necessities as social infrastructure, and nature, heritage, and local food as rural attractions.

(%)

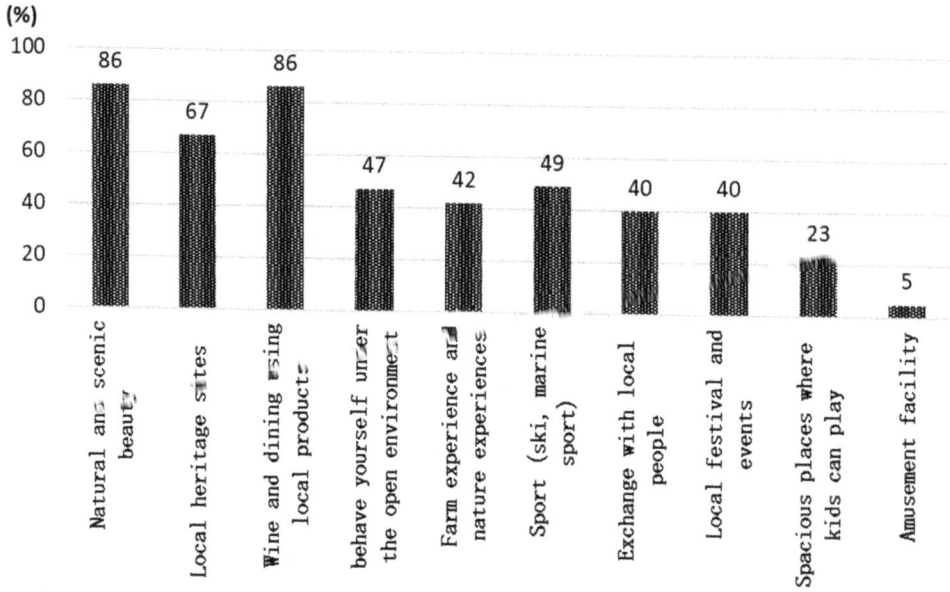

Figure 4: Necessary additional values for workcation/telework (multiple choice). *(Source: Internet questionnaire survey to three Metropolitan inhabitants conducted by Hyakusen-Renma in June 2020.)*

(%)

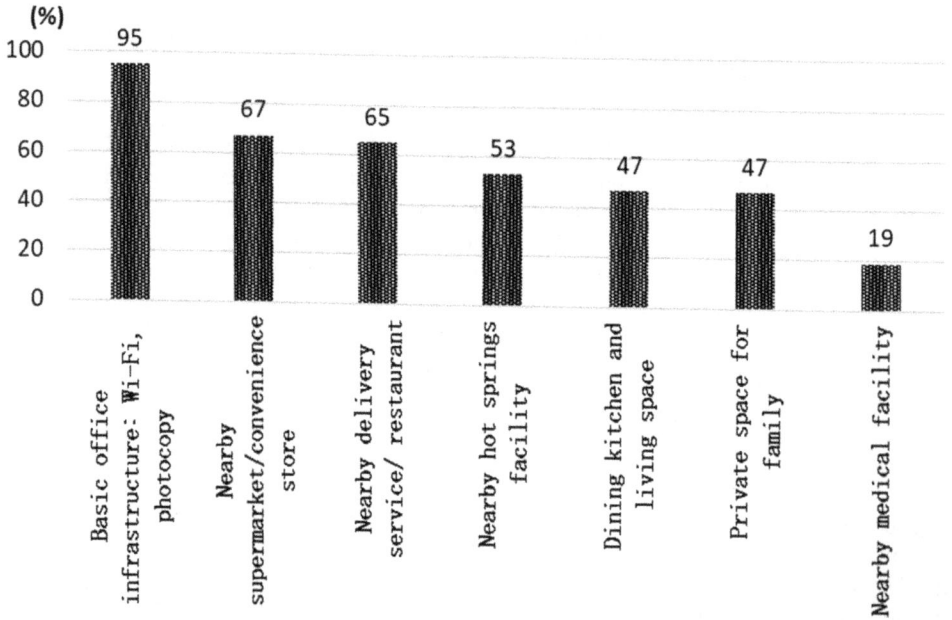

Figure 5: Necessary facility/equipment for workcation/telework (multiple choice). *(Source: Internet questionnaire survey to three Metropolitan inhabitants conducted by Hyakusen-Renma in June 2020.)*

## 4 EXPANDING DOMAIN OF HOSPITALITY: E-HOSPITALITY

To put these potential demands into practice as a rural-tourism business, the author considers what was not included in the survey results, which were hospitality issues. An OTA that is specialized in SIT, such as Stay Japan, has been emerging in Japan. As OTAs have diffused, the service domain that they provide has expanded from a basic accommodation-booking function to the provision of information on nearby attractions, transportation booking, etc.

Thus, let me consider what influence the development of OTAs exerts on what hospitality should be, which is termed as one of "the e-hospitality effects". OTA services are convenient for tourists and increases the number of users due to network externality. This expansion further stimulates increases in the types and content of OTA services. It should be noted that the e-hospitality effects are not limited to the improvement of tourists' convenience. Although hospitality issues have not been studied in the rural tourism arena, it is important for managers to appropriately understand the effects of e-hospitality [5], [13].

In the field of hospitality management and marketing, Wood touched upon the Ritzer's well-known concept of McDonaldization [14] in hospitality industry as service industry. This concept represents the principles of production rationalization applied to service industry originated from manufacturing industry [15]. This is one direction of hospitality. The digital transformation (DX) is also considered as the same line of evolution [15].

Bojanic and Reid pointed out the effect of improved customer communication and relations in electronic commerce from marketing perspectives [16, pp. 229–244]. Bojanic and Reid also deal with social media from advertising point of view [16, pp. 274–294].

Busulwa et al. addresses the digital disruptions such as disruptions in available data and disruptions in competitive landscape caused by digital transformation [17]. Thus, how to cope with these digital disruptions are crucial issues to the digital transformation for rural tourism operators as well [17].

In keeping that hospitality issue in mind, the author considers the significance of e-hospitality in connection with traditional face-to-face hospitality (Fig. 6). E-hospitality means that hospitality services are provided in a non-contact way, i.e., the internet. The components of services are booking, transportation, trip-planning support, selling local products, exchange with local people, training programs, etc. [5]. Fig. 6 contrasts the two phases before (on the left) and after COVID-19 (on the right). The difference between the two phases is that the domain of e-hospitality enlarged after COVID-19. Traditional face-to-face hospitality is risky for tourists in the sense that the quality of service is uncertain until experienced, which is a feature of experience goods.

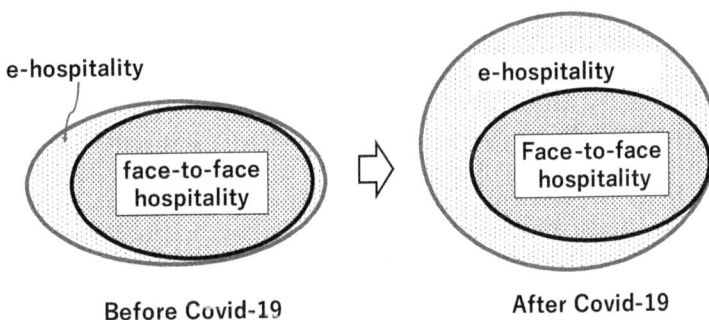

Figure 6: Conceptual model of hospitality business before/after COVID-19. *(Source: Modified based on [13])*

When e-hospitality service improves, to some extent tourists can grasp the quality of hospitality at the destination before arrival and uncertainty for tourists is reduced. The risk-easing benefits are not limited to tourists but also to rural tourism managers. Because managers can convey service contents beforehand, which can reduce the risk of receiving complaints from tourists who are disappointed due to unreasonably high expectations. This is the second e-hospitality effect generated for both tourists and managers of rural tourism.

This means that part of the hospitality is already implemented on the internet screen rather than on-site at the gate of accommodations. Thus, managers should recognize that hospitality starts on the website, which is what is different from face-to-face hospitality. Since the domain of e-hospitality will continue to expand further, it will become crucial for managers to properly understand this point.

From the above consideration, we can say that face-to-face hospitality and e-hospitality are substitutable. COVID-19 rapidly promotes substitution of face-to-face hospitality with e-hospitality. Although the substitutable area is not small, there is no doubt that essential elements belong to face-to-face hospitality. In this respect, complete substitution is not possible because face-to-face hospitality is indispensable to determine the final satisfaction of tourists. To put it another way, this substitution enables managers to save time and energy and put this saved time and energy into enhancement of the core hospitality provided face-to-face. This is the third e-hospitality effect.

Thus, it is important for effective marketing and management of rural tourism to integrate e-hospitality and real face-to-face hospitality since e-hospitality will continue to expand its domain through the substitution and development of its own new area.

To summarize, the new normal creates opportunities for the rapid progress of e-hospitality and increase its importance, such as improvement of convenience, risk reduction and saving time and energy for face-to-face hospitality. Thus, it will be increasingly necessary to integrate e- and face-to-face hospitality, which is termed here as integrated hospitality. Since the issue of integrated hospitality has been little investigated, this will be a new research topic under the new normal. Consequently, it will become necessary to consider policy design of support measures on how to properly manage integrated hospitality for rural tourism in the future.

## 5 CONCLUSION

Rural tourism in Japan has been far behind the national trend of inbound tourism from abroad. Although COVID-19 exerted a serious reduction in the number of stays in rural tourism, i.e., loss of one-third of the market compared with pre-COVID-19 in 2019, marginal dependence on the inbound market avoided devastating consequences. During the slow business period of COVID-19, increases in the number of meal menus and progress in using OTAs were observed, the continuation of which will be effective for the evolution of rural tourism after COVID-19.

On the positive side of COVID-19, although COVID-19 shrunk the B-to-B demand for farm and fishery products, it also generates opportunities to let rural tourism managers rediscover rich local resources and reconfirm the local identity, which could lead to new rural tourism businesses. Thus, this paper revealed that new demands such as for micro-tourism and workcation are emerging based on the internet questionnaire survey to metropolitan inhabitants in Japan. The time-flexible mature generation prefers micro-tourism while workcation is preferred by younger generations who seek a flexible work style. This shows that the two new demands are segmented markets. The significance of workcation is a new lifestyle away from the congested risky urban lifestyle under the new normal, which is a novel perspective never observed in the former rural-tourism programs. Thus, workcation can open the door to rural tourism for the young generation that has had little interest in rural tourism before.

From a hospitality perspective, the domain of face-to-face hospitality has partially moved to non-contact e-hospitality. This partial transition means not only simply changing from face-to-face to online, but also represents changes as to where and when hospitality is begun to be offered, i.e., from starting on site to starting on the website. The rural tourism manager should adequately understand this shift caused by surging e-hospitality. The shift from face-to-face to e-hospitality will deepen due to rapid IT progress.

Nevertheless, this trend does not mean that the role of face-to-face hospitality will shrink but that a chance to enhance the original value of face-to-face hospitality will be offered. In this context, it is essential to explore how to maximize the value of face-to-face hospitality in taking advantage of e-hospitality. Consequently, we should focus more on how to offer integrated hospitality to tourists and design support measures for that purpose. COVID-19 also makes us realize that we need to launch full-scale hospitality research in rural tourism.

## ACKNOWLEDGEMENT

This study was financed by Grant-in-Aid for Scientific Research (Kakenhi) No.18H03965, No. 20H04444, and No. 20H03086 by Japan Society for the Promotion of Science (JSPS).

## REFERENCES

[1]  Fotiadis, A., Polyzos, S. & Huan, T.C.T., The good, the bad and the ugly on COVID-19 tourism recovery. *Annals of Tourism Research*, **87**, 103117, 2021. DOI: 10.1016/j.annals.2020.103117.

[2]  Sharma, G.D., Thomas, A. & Paul, J., Reviving tourism industry post-COVID-19: A resilience-based framework. *Tourism Management Perspectives*, **37**, 100786, 2021. DOI: 10.1016/j.tmp.2020.100786.

[3]  Škare, M., Soriano, D.R. & Porada-Rochoń, M., Impact of COVID-19 on the travel and tourism industry. *Technological Forecasting and Social Change*, **163**, 120469, 2021. DOI: 10.1016/j.techfore.2020.120469.

[4]  Qiu, R.T.R., Wu, D.C., Dropsy, V., Petit, S., Pratt, S. & Ohe, Y., Visitor arrivals forecasts amid COVID-19: A perspective from the Asia and Pacific team. *Annals of Tourism Research,* **88**, pp. 1–16, 2021. DOI: 10.1016/j.annals.2021.103155.

[5]  Ohe, Y., Exploring a way forward for rural tourism after the Corona pandemic. *Journal of Global Tourism Research*, **5**(2), pp. 1–2, 2020. DOI: 10.37020/jgtr.5.2_105.

[6]  Polukhina, A., Sheresheva, M., Efremova, M., Suranova, O., Agalakova, O. & Antonov-Ovseenko, A., The concept of sustainable rural tourism development in the face of COVID-19 crisis: Evidence from Russia. *Journal of Risk and Financial Management*, **14**(1), p. 38, 2021. DOI: 10.3390/jrfm14010038.

[7]  Ohe, Y., *Quantitative Economic Approaches to Rural Tourism*, Agriculture and Forestry Statistics Publishing: Tokyo, 2017. (In Japanese.)

[8]  Ohe, Y., *Community-Based Rural Tourism: A Microeconomic Approach*, Springer: Singapore, 2020. DOI: 10.1007/978-981-15-0383-2.

[9]  Ministry of Agriculture, Forestry and Fisheries, Japan, On the Situation of Nohaku Version at 7th January, 2021. (In Japanese.) https://www.maff.go.jp/tokai/noson/keikaku/nouhaku/attach/pdf/index-1.pdf. Accessed on: 15 Jan. 2022.

[10]  Ohe, Y., From product-out to market-in rural tourism. *HortResearch*, **73**, pp. 8–9, 2019. DOI: 10.20776/S18808824-73-P8.

[11] Ohe, Y., Overcoming challenges toward sustainable rural tourism: A perspective of community-based tourism. *Journal of Rural Planning Studies*, **38**(1), pp. 10–14, 2019. (In Japanese.) DOI: 10.2750/arp.38.10.

[12] Azuma, T., From inbound-tourism market to domestic tourism market and workcation: Can tourism in Japan get back her feet? *Chuokoron*, pp. 136–143, 2021. (In Japanese.)

[13] Ohe, Y., Exploring new directions of rural tourism under the new normal: Micro-tourism and workcation. *Japanese Journal of Tourism Studies*, **20**, pp. 1–9, 2022. (In Japanese.)

[14] Wood, R.C., *Hospitality Management: A Brief Introduction*, SAGE: London, pp. 6–9, 2015.

[15] Ritzer, G., *The McDonaldization of Society: Into the Digital Age*, 9th ed., SAGE: Thousand Oaks, CA, 2019.

[16] Bojanic, D.C. & Reid, R.D., *Hospitality Marketing Management*, 6th ed., John Wiley: Hoboken, NJ, 2017.

[17] Busulwa, R., Evans, N., Oh, A. & Kang, M., *Hospitality Management and Digital Transformation: Balancing Efficiency, Agility and Guest Experience in the Era of Disruption*, Routledge: Abingdon, pp. 43–50, 2021.

# SECTION 3
# CULTURAL TOURISM

# NEW PARADIGM OF SPIRITUAL TOURISM: ADDING AN IMPORTANT LAYER TO SUSTAINABLE TOURISM

LORENZ POGGENDORF
Department of International Tourism Studies, Toyo University, Japan

## ABSTRACT

Sustainable tourism has been on the rise for many years. This paper argues for a connection between sustainable tourism and spiritual tourism, and moreover explains the similarities and differences between traditional spiritual tourism and new spiritual tourism by means of various case studies. The main goal of the paper is, however, to illustrate the potential of new forms of spiritual tourism for an expanded understanding of sustainability. Until now, sustainable tourism has been focussing on balancing environmental, social and economic issues in the external material world. To these important goals, tourism should add the goal of a sustainable trip in the sense of long-term benefits towards subjective wellbeing beyond the trip itself. This is because ultimately, we will only achieve balance in the outer world through balance in our inner world. By adding spiritual tourism to the concept of sustainable tourism, we will be able to integrate the spiritual dimension of human existence into an even more fulfilling travel experience.
*Keywords: spiritual tourism, sustainable tourism, classification, case studies, potential.*

## 1 INTRODUCTION

What is spiritual tourism? First of all, it needs to be said, there is no universally valid definition of "spiritual tourism". Each person experiences life, spirituality, and travel differently. The reason for this is that each of us has a diverse background and personality, shaped by our family, education, profession, culture, and nation from which we come.

This being said, there exist various definitions through previous studies. In this paper, firstly, the author would like to build on this and point out two fundamental trends in spiritual tourism. Secondly, we will explain the basic perspective and relevance of this work.

### 1.1 Definition of terms

"Spiritual tourism" stands for self-discovery and wellbeing maintenance as well as seeking healing for the soul [1]. It embodies a holistic approach that combines and, at best, harmonises the external physical journey with an inner journey to explore ourselves.

Smith and Kelly [2] use "holistic tourism" as an umbrella term, to which they assign spiritual and other forms of tourism. They not only define spiritual and religious tourism, but also classify wellness tourism and alternative healing practices (such as spa treatments, herbal medicine, yoga, massage, acupuncture, etc.) as holistic tourism [2].

Liutikas [3], analysing the underlying values of pilgrims, uses "value-based tourism" as a generic term and presents definitions of various subtypes, including "spiritual tourism" and "pilgrimage tourism" [3].

Each of these definitions has its justification. However, at the same time it becomes clear that a variety of terms and interpretations are used. For this reason, it is natural for every author to use and clarify his own terminology.

In this paper we will use the terms "traditional spiritual tourism" and "new spiritual tourism".

WIT Transactions on Ecology and the Environment, Vol 256, © 2022 WIT Press
www.witpress.com, ISSN 1743-3541 (on-line)
doi:10.2495/ST220061

"Traditional spiritual tourism" refers to pilgrimage tourism and religious tourism – approaching and visiting famous religious spots, such as churches, monasteries, shrines, temples or mosques. Its focus is on cultivating prayer and faith community, and on facilitating collective worship for followers of a certain religion. Traditional spiritual tourism in the form of journeys and pilgrimages to celebrated places of faith is a centuries-old tradition which continues throughout the world [4].

"New spiritual tourism" is based on recent trends in spirituality, promoting practices such as mindfulness and a personal path that seeks happiness within. It takes place at unique destinations in beautiful landscapes, including "power spots", and often includes exercises for spiritual development, such as yoga, qi-gong, meditation, and breathwork retreats [5].

These definitions serve as a point of reference. A clear separation is not always possible, because there are also overlaps, and people of various backgrounds and personalities as well as trips cannot always be divided into two groups of "traditional" and "new". Still, both terms have in common that visitors who join spiritual tourism activities are motivated to connect to a "higher source".

## 1.2 Objectives

Through spiritual tourism, is it possible to provide a travel experience that touches the hearts of travellers beyond the journey itself? What unites and differentiates traditional and new spiritual tourism? A sustainable trip in the sense of long-term benefits should have a positive impact on travellers' emotional state and wellbeing, even after returning to everyday life. The central argument of this paper is that we should look at concepts that combine both external happiness and internal happiness.

### 1.2.1 External happiness
Modern travel, such as classical sightseeing or beach tourism, mainly focusses on external happiness. In other words, it's primarily about satisfying elementary physical and emotional needs through providing various distractions, events, and amenities. There is nothing wrong with this. Maslow already points out that all people want to have their basic physical and emotional needs met [6]. However, we also know that these consumption-oriented forms of travel often lead to serious problems in the environmental and social spheres [7], [8].

In addition, it can be assumed that tours focussing on external features will not always bring continued fulfilment and satisfaction. In daily life, modern people spend a lot of time watching news and advertisements, reading messages and shopping online using media such as the Internet, SNS, and smartphones, and this behaviour or lifestyle naturally continues on the journey. As convenient as our smartphones may be, they also constantly distract us from essential things. As a consequence, often, we don't find time to connect with ourselves and others at heart level [9]. That is why this paper discusses more conscious travel experiences.

### 1.2.2 Internal happiness
What we feel inside our heart is key. No matter how perfectly planned a visit to a tourist destination might be, whether it really turns out to be a meaningful and memorable experience one ultimately depends on how one experiences it in one's own heart.

The current idea of tourism, as a service industry, is to make people happy during the trip, by providing amenities and comfort to the greatest extent possible for those who can spend money. It is a hospitality culture, especially in Japan. Of course, it's also a way to make people happy, through having a good time, but all of this is superficial. And it is no exaggeration to say that such a mindset directs modern society as a whole.

However, people are beginning to look more and more at their inner happiness. Since the mid-20th century, in large part through the influence of Asian spiritual leader such as Suzuki, Eastern thought has become increasingly popular in the West, and has been especially influential upon New Age culture [10]. In the last few years, we have observed a global trend of "mindfulness" towards more internal happiness through practices such as meditation, yoga, and qigong. Picking up on such trends and considering their future potential in spiritual tourism is one goal of this paper.

### 1.3 The connection between sustainable tourism and spiritual tourism

Recently, the author outlined the evolution of "sustainable tourism" or "green tourism" [11]. In this paper, he would like to emphasise that both sustainable tourism and spiritual tourism – despite differences – also have a common value base, for both are concerned with safeguarding the world and creating a brighter future.

For sustainable tourism, the focus is on using natural and cultural resources in better ways, in order to protect and harmonize the environment, local enterprises, and culture in the long run.

For spiritual tourism, the focus is on using one's inner resources for a more sophisticated and peaceful existence. If we manage to create more harmony and peace in our inner being, we will be able to use this power for the benefit of our fellow men and the environment in the outer world.

We are not just a body made of flesh, bones, and blood. Human existence combines physical, emotional, and spiritual aspects. The latter is often neglected, both in daily life and in conventional tourism.

The late Austrian physician and psychiatrist Professor Dr Viktor Frankl (1905–1997) founded logotherapy and existential analysis, which is often referred to as the "Third Viennese School of Psychotherapy". In decades of research into human fulfilment and the meaning of life, he emphasizes the spiritual aspect of human beings the most. Frankl sees human existence basically on three levels: (1) body (physiological functions), (2) mind (emotion and thought), and (3) spirit (higher meaning, our calling, service to the world) [12]. His explanatory model of human reality, simple at first glance but profound in its implications, may also help to understand the role and potential of spiritual tourism. Only when we connect with our spiritual core, can we lead a truly fulfilled life and serve the world as the best version of ourselves. And spiritual tourism can help us with achieving that goal.

Therefore, our task is not only to meet external needs (of nature and the environment, tourists, enterprises, and local people) by sustainable tourism but also internal needs (inner peace and harmony). By adding spiritual tourism to the concept of sustainable tourism, we will be able to integrate the spiritual dimension of human existence into an even more fulfilling travel experience.

### 1.4 Methodology

This paper is based on literature studies, site visits, qualitative interviews, and email enquiries.

In the introduction, reference was made to some common definitions of spiritual tourism, and basic views and approaches to the topic were presented in a condensed form.

Based on this, the first part of this paper builds on previous studies by the author about "traditional spiritual tourism". To create his own picture through field research, the author has dealt with famous pilgrimage sites in Austria and Japan, where he conducted in-depth interviews with leaders of the respective places of worship, about their identity, and about how they deal with tourists in these sacred places.

The second part of this paper deals with the phenomenon of "new spiritual tourism". Two case studies illustrate that this form of spiritual tourism has different objectives and is implemented differently. Due to the travel restrictions of the corona pandemic, field research has been difficult to conduct. However, knowledge gaps were filled by online research and email enquiries as much as possible.

The structure and method of this work is summarised in Table 1.

Table 1: Methodology overview.

| Step | Content and structure | Sources of evidence (data type) |
|------|----------------------|--------------------------------|
| 1 | Introduction to the topic and definitions of terms | • Existing studies concerning spiritual tourism and its search for meaning in life (secondary data) |
| 2 | **Traditional spiritual tourism:** Interviews about values and tourist positioning of the Catholic monastery Benedektinerstift Göttweig (WH) in the Wachau county, Austria, and Grand Shrine of Nachi (WH), Kumano Sanzan area, Japan | • In-depth interviews with key persons of both destinations • Site inspections by the author (primary data) |
| 3 | **New spiritual tourism:** Online research and enquiries about two case studies – The "Meditation House" of Hotel "Das Kranzbach" in Germany and the Aura-Soma Centre "Dev Aura" with its accompanied "Shire Farm" that offer corresponding programs | • Online research and email enquiries • Site inspection by the author of one of the sites (secondary and primary data) |

## 2 RESULTS

### 2.1 Traditional spiritual tourism – pilgrimage to religious spots

In the following we would like to summarise the results of previous studies about "traditional spiritual tourism" [11], [13]. The original aim of these studies was to reflect on the essential importance of religious spots and to draw conclusions for quality heritage tourism. However, they also reveal insights in comparison to new forms of spiritual tourism.

In Austria, investigations of the impressive World Heritage and Catholic Monastery "Benedektinerstift Göttweig" on the top of a hill next to the River Danube showed that through open devotions, especially midday prayer, visitors can directly experience the spirituality of the monastery (Fig. 1). There are both day visitors for sightseeing and those who stay overnight, some of whom are guests at the monastery and take part in services with the monks. This means that the visitor is free to decide for himself how far he wants to engage with the monastery.

Keeping with the ancient tradition of the Benedictine Order, Stift Göttweig has its own inn and restaurant, which is able to impress its visitors with fresh regional ingredients and

Figure 1:  Benedictine Abbey Stift Göttweig in an outstanding rural setting near the River Danube. © Stift Göttweig/OEBH.

sustainable management. In addition, Stift Götttweig is connected to pilgrimage routes via the Austrian Way of St. James and other long-distance hiking trails that lead all the way to the world-famous Spanish Camino de Santiago.

Next, in Japan, the "Grand Shrine of Nachi" (*Kumano Nachi Taisha*) is part of the Kumano Sanzan pilgrimage network in Wakayama Prefecture. Being also a World Heritage Site, it represents a unique spot of ancient syncretic faith (of Shintōism and Buddhism) combined with an old mountain worship called *Shugendō*. Surrounded by breath-taking natural landscape, with its sacred Nachi Waterfall, which is the highest drop in Japan, this outstanding religious site attracts visitors from all over the world (Figs 2 and 3).

The shrine and its sacred waterfall attract three groups of visitors. The first group come only as tourists, on account of the waterfall and the World Heritage status. The second group are believers who worship the twelve gods of Kumano. The third group are those who worship the sacred Nachi waterfall itself. Sometimes, it also often happens that visitors spontaneously develop a feeling or special relationship with this place. At first, they come just to see the waterfall, but then they are completely moved by what they feel at the waterfall and become believers. Or they observe other visitors praying and thus open themselves to the gods (*kami*) of this place.

The Great Shrine of Nachi is part of an extensive pilgrimage network that runs through the lush Kii Mountains and includes pilgrimage trails connecting Kumano Sanzan with the world-famous Buddhist temple complex "Koyasan" and with the Shintō "Grand Shrine of Ise".

The Kumano Sanzan pilgrimage routes in Japan and Santiago de Compostela in Spain, both World Heritage Sites, also have close cooperation as partner networks.

Figure 2: The Daimonzaka Nakahechi pilgrimage trail, leading to the Kumano Nachi Taisha Grand Shrine and Nachi Falls.

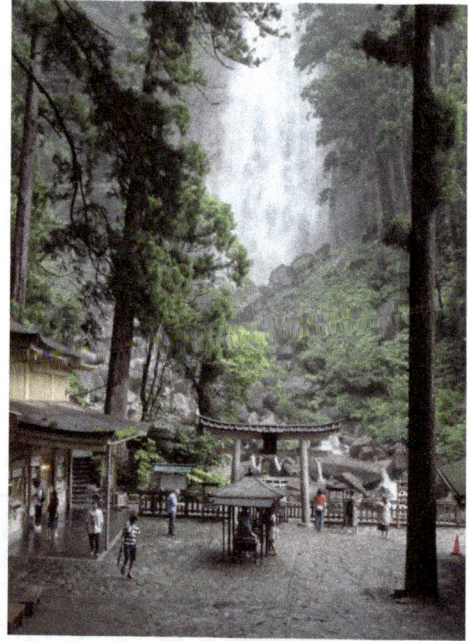

Figure 3: The sacred Nachi Falls of Kumano Nachi Taisha Grand Shrine.

Such destinations, which have grown over centuries, possess their own special aura and attraction. By preserving their beliefs and authenticity, they also meet the strict criteria of a world heritage site. These are good examples of places that have a strong connection to traditional spiritual tourism. They keep their long tradition, but also open up to new visitors from abroad. The long history, the beauty of the place, the clear positioning in terms of content and religion, and the legal protection status of such sites – all such aspects together make this outstanding sense of place possible.

## 2.2 New spiritual tourism – beauty and tranquillity inside and out

Traditional spiritual tourism has been taking place for centuries. The previous examples have briefly shown that it is the context of historical significance, natural setting, and a living faith and pilgrimage tradition that creates the splendour and unique attraction of such places.

With new spiritual tourism, things look a little different. In most cases, travel destinations do not have the same level of historical, cultural, and natural resources. Of course, for new spiritual tourism, the location and quality of place also play an important role, but most of the time the level of World Heritage is not reached. However, it is not essential to keep up with traditional World Heritage. The decisive factor is rather to combine the quality of the given place and environment (outer quality) with the quality of the respective course or activity offers for inner growth (inner quality) in the best possible way.

In the following, two examples are presented that offer new forms of spiritual tourism – beyond religion and denomination.

The first example is the "Meditation House" of the wellness hotel "Das Kranzbach", Bavaria, Germany. The second example is the Aura-Soma Centre "Dev Aura" and its "Shire Farm" in Lincolnshire, England, United Kingdom. Both places offer contemplation, tranquillity, relaxation and introspection, in charming rural surroundings.

2.2.1  Case study 1: "Das Kranzbach Meditation House", Bavaria, Germany

"Wir wollten für unsere Gäste immer einen besonderen Ruheort schaffen".
[*We always wanted to create a special resting place for our guests.*]
Dr. Jakob Edinger (owner of the Hotel Das Kranzbach)

The Hotel "Das Kranzbach" is one of the leading high-end wellness hotels in Germany. It is located in the Bavarian Alps near the Austrian border, surrounded by appealing natural phenomena such as alpine meadows, pristine mountain peaks, and deep forests (Fig. 4).

Figure 4:  Hotel "Das Kranzbach", Bavaria, Germany, in a stunning alpine landscape setting. © Das Kranzbach.

The object of this study is the "Meditation House", which was designed by the Japanese star architect Kengo Kuma and built in 2018 [14]. The meditation house is separated from the hotel building complex on purpose – one has to walk a footpath up the forested hill to enter it.

In 2016, Kengo Kuma received the Global Award for Sustainable Architecture [15]. High environmental standards are very important to him. Correspondingly, the construction of the "Meditation House" followed strict environmental standards and – to a certain degree – Japanese design. For instance, only as many trees as necessary were cleared and the trunks were pulled out of the forest individually by horse to protect the soil against compaction. In addition, for the posts and beams, the architect used regional wood from native fir trees (Fig. 5).

WIT Transactions on Ecology and the Environment, Vol 256, © 2022 WIT Press
www.witpress.com, ISSN 1743-3541 (on-line)

Figure 5:    The Meditation House, designed by the Japanese star architect Kengo Kuma, inside to outside view – "to feel one with nature". © Das Kranzbach.

As the name suggests, the "Meditation House" is a resource for retreat and relaxation. It has three major purposes:

1.    A space of silence in the forest
2.    To feel one with nature and find peace of mind
3.    To offer classes for meditation, yoga, and qi-gong

Further, the hotel owner emphasises the following features of the meditation house: this is not a fashion but a long-term project. It creates a unique sales point. The demand among guests for "inner relaxation" is becoming more important for an increasing number of guests year after year. This includes the desire for nature, peace and silence, and for a retreat during which to ground oneself again. Before the start of the Corona pandemic, the hotel "Das Kranzbach" had an annual average of about 200 guests per day in the hotel, about 15% of whom took part in daily yoga and meditation courses. This trend is increasing. The yoga and meditation offers are decisive in encouraging these guests to book. Taster guests make up another 5%. Following an easing of the pandemic, this positive trend is expected to strengthen further [16].

The retreat concept is successful for two reasons. First, in our challenging times, many people need effective tools for relaxation and introspection, and second, its unique setting in nature provides the best opportunities for a retreat.

### 2.2.2 Case study 2: "Dev Aura" and "Shire Farm" of Aura-Soma Ltd. in Lincolnshire, England

"Der größere Lehrer ist in dir selbst. Was wir anbieten, sind nur Anhaltspunkte".
[*The greater teacher is within yourself. What we have to offer are only clues.*]
Vicky Wall (founder of Aura-Soma) [17]

Aura-Soma is a natural, holistic colour therapy from England, which aims to bring people back into inner balance [18].

"The Aura-Soma® Colour Care System is a non-intrusive, self-selective colour system of great beauty which offers you the opportunity for awareness and transformation."
The Aura-Soma Academy

The core of Aura-Soma is made up of around 120 "Equilibrium Bottles". They are combinations of natural plant oils, high-quality essential oils, healing herbs, and crystal energies. Each bottle consists of 50% plant oil and 50% ultrapure water, each with its own colour, and is blended into an emulsion that is applied to the body on the skin.

The main idea of Aura-Soma is that each person intuitively chooses colour combinations based on their personality, which best support them in their personal spiritual development. The colours chosen are also related to the colours of the Indian chakra system or energy centres of the body, and are believed to be capable of balancing and revitalizing them. Further therapeutic Aura-Soma products called "Pomander" and "Quintessence" have an even more subtle effect on the human energy system.

"The products have an energetic vibration that goes beyond words".
The Aura-Soma Academy
"You are the colours you choose".
Vicky Wall

For the training in Aura-Soma colour therapy, Aura-Soma Ltd. has its own seminar house with guest rooms and a special garden, called "Dev Aura" and managed by "The Academy" (Fig. 6). This beautiful training centre has developed continuously over the past 30 years. Meanwhile, visitors arrive from all over the world, on average about 400–500 annually. Only COVID-19 recently interrupted the influx [19].

Figure 6:    Lecture room of the Aura-Soma Academy at "Dev Aura" surrounded by an extraordinary garden. © Aura-Soma Products Limited.

Aura-Soma Ltd. also has its own fields, the "Shire-Farm". Here, most of the medicinal herbs needed for their products are grown. The cultivation is biodynamic – in harmony with nature and natural rhythms – and therefore absolutely sustainable. The author had the opportunity to visit both the seminar house "Dev Aura" and the "Shire Farm" years ago and was very positively impressed.

Both the seminar house of the Aura-Soma Academy "Dev Aura" and the organic "Shire-Farm" offer a platform for mainly three purposes:

1.  Courses about the Aura-Soma colour therapy and exercises in the seminar house, such as silent meditations indoors, for spiritual introspection
2.  Meditative walks through the fields, gardens, and woods of Shire-Farm, to get into deeper contact with Mother Earth
3.  Special excursions to "sacred landscapes" (power spots), for instance, to the mythical spot of Glastonbury, called the "Sacred Tour – Glastonbury Experience".

## 3  DISCUSSION AND CONCLUSION

### 3.1  Achievements and limitations

This paper looked at tourism in the context not only of material sustainability (related to environmental, economic, and social aspects) but also immaterial sustainability (related to a lasting sense of wellbeing and personal development). Spiritual tourism aims at a happiness that we can discover and feel deep within us – and that goes beyond common physical comforts, pleasures, and luxuries.

The case studies about traditional spiritual tourism of the monastery in Austria and Shinto shrine in Japan have shown that visitors to such World Heritage destinations vary from mere sightseeing to spontaneous participation in prayers or ceremonies to deep devotion.

They are first attracted by the fame of the place and its unique scenic location. In the case of the Monastery Stift Göttweig, it is the impressive building complex and its setting on the mountain overlooking the River Danube and the vineyards of the Wachau. In the case of the Great Shrine of Nachi, it is the sacred Nachi waterfall in stunning nature, together with the shrine and temple complex and its pilgrimage network in a remote mountainous area.

At the same time, their attraction also relates to the invisible, unprovable and yet tangible uniqueness of these places, its connection to a higher, divine source.

The case studies about new spiritual tourism, with the meditation house in Germany and Aura-Soma centre in England, reflect new trends towards relaxation and introspection – in the face of a modern performance society increasingly plagued by haste and stress. Here, too, places are visited in the midst of rural beauty. However, the focus is not on prayers at a spot of glorious history or with World Heritage status. Rather, the primary aim is to immerse oneself in meditative silence and explore oneself in an undisturbed place under guidance.

I should hereby observe that this study is not about judging which form of spiritual tourism is better. It is rather about broadening understanding and pointing out new possibilities.

The case studies illustrated in this paper provide some initial insights into spiritual tourism, its core characteristics, and its multifaceted offerings.

What is missing, however, is data on the number and quality of ongoing services and the elaboration of success criteria. What is a truly authentic offer of high quality? For the traveller and seeker, various course offerings and destinations make it difficult to choose. Whether yoga, qi gong, naturopathic therapies combined with wellness and self-awareness – everyone claims to have the best offer. The present work is therefore merely an introduction to the subject and a stimulus to look at the meaning and purpose of travel with new eyes.

Unlike in classical tourism, where benefits and service offers can be defined more easily, there are no guaranteed results with spiritual course offers. However, there are no guaranteed results when you visit the medical doctor, either. Therefore, no false expectations should be created, but offers should always be communicated openly and fairly.

## 3.2 Implications for further research

So far, the author has only interviewed the supplier side. The survey of travellers and course visitors in the field of new spiritual tourism is still pending. Another way could be to examine online feedback, such as blogs and comments from corresponding participants.

A further aspect meriting future research is the development of quality criteria in spiritual tourism.

## 3.3 Implications for practice

For new spiritual tourism, a crucial point is to create offers that enable the visitor to experience and feel the inner self beyond the boundaries of religious denominations. It is all about allowing people to break out of a restless everyday life and come back to true contemplation. This is because deep down we are searching for meaning, which is becoming more and more difficult to find in today's times of permanent media exposure and the expectation of being approachable and available at all times.

An important question for all of us is, what is our main goal, distraction or deepening our sense of life? For those who are interested in deepening themselves, we should offer new travel opportunities that dig deeper into our own hearts. However, such offers can only be developed by or with those who have already walked this path inwards. Finding truly competent spiritual guides who are primarily concerned with the matter at hand and less with fame or business remains an ongoing task.

In practical terms, to initiate an inner change that has a positive effect beyond the duration of the trip, it is recommended to combine lessons learned on-site with corresponding online products (apps) to support the continuation of learned meditation practice or physical exercises at home.

After all, sustainable is only that which we can successfully incorporate into our everyday lives and deepen step by step. If we succeed in this, the outer journey also becomes an inner journey. The aim of this essay was to provide some first hints for such a journey.

## ACKNOWLEDGEMENTS

The author would like to express sincere gratitude to everyone involved in his research for this paper. In particular, he thanks Prior Maximilian and Ms. Eveline Gruber-Jansen, Head of Tourism and Culture, of the Monastery Stift Göttweig, Mr. Hermann Paschinger, former director of the marketing platform "Klösterreich", and Shintō-priest Mr. Itoh from the Kumano Nachi Taisha Grand Shrine for granting in-depth interviews.

Sincere thanks also go to Dr. Jakob Edinger, owner of the Hotel Das Kranzbach, and Ms. Sachie Shinohara of Aura-Soma Ltd. for their wonderful support. As always, Dr. Alice Freeman, Oxford, skilfully helped with correcting English mistakes. I am indebted to them (and to all those not mentioned here). Finally, permission to print has been obtained for all photos not taken by the author.

## REFERENCES
[1]   Norman, A., The varieties of the spiritual tourist experience. *Literature and Aesthetics*, **22**(1), pp. 20–37, 2012.

[2]     Smith, M. & Kelly, C., Holistic tourism: Journeys of the self? *Tourism Recreation Research*, **31**(1), pp. 15–24, 2006. DOI: 10.1080/02508281.2006.11081243.
[3]     Liutikas, D., The manifestation of values and identity in travelling: The social engagement of pilgrimage. *Tourism Management Perspectives*, **24**, pp. 217–224, 2017. DOI: 10.1016/j.tmp.2017.07.014.
[4]     Raij, R. & Griffin, K. (eds), *Religious Tourism and Pilgrimage Management: An International Perspective*, 2nd ed., CAB International, 2015.
[5]     Smith, M., Holistic holidays: Tourism and the reconciliation of body, mind and spirit. *Tourism Recreation Research*, **28**(1), pp. 103–108, 2003. DOI: 10.1080/02508281.2003.11081392.
[6]     Maslow, A.H., A theory of human motivation. *Psychological Review*, **50**, pp. 370–396, 1943.
[7]     Weaver, D.B., Ecotourism in the context of other tourism types. *The Encyclopaedia of Ecotourism*, CABI Publishing, 2001.
[8]     Honey, M., *Ecotourism and Sustainable Development*, 2nd ed., Island Press: Washington, DC, 2008.
[9]     Varchetta, M., Fraschetti, A., Mari, E. & Giannini, A.M., Social media addiction, fear of missing out (FoMO) and online vulnerability in university students. *Revista Digital de Investigación en Docencia Universitaria*, **14**(1), e1187, 2020. https://revistas.upc.edu.pe/index.php/docencia/article/view/1187/1111. Accessed on: 28 Mar. 2022.
[10]    Suzuki, D.T., *An Introduction to Zen Buddhism*, Eastern Buddhist Society: Kyoto and Evergreen Black Cat Book: New York, 1934 and 1964.
[11]    Poggendorf, L., Travel for sightseeing or for an enriching life? A reflection on value-based rural heritage tourism in Austria and Germany. *Toyo University Graduate School Bulletin*, **58**, pp. 61–100, 2021.
[12]    Frankl, V., Der Mensch auf der Suche nach Sinn. Zur Rehumanisierung der Psychotherpie (Man in Search of Meaning). Herderbücherei, 1959 and 1975.
[13]    Poggendorf, L., Sacred sightseeing spots in Japan: Comparison of two prominent shrines and their local tourism policies (Nihon no shinseina kankō supotto — futatsu no chomeina jinja to sono chihō kankō seisaku no hikaku). *Contemporary Social Studies (Gendai Shakai Kenkyū)*, **16**, pp. 111–121, 2019.
[14]    Das Kranzbach, Meditation in Kranzbach, Bavaria, Germany, 2022. https://www.daskranzbach.de/de/meditation-in-kranzbach.html. Accessed on: 27 Mar. 2022.
[15]    The Cité, Global award for sustainable architecture, 2022. https://www.citedelarchitecture.fr/en/article/global-award-sustainable-architecture. Accessed on: 28 Mar. 2022.
[16]    Edinger, J., About the Meditation House. Answering the author's questions via email reply on 5 Mar. 2021 (owner of the Hotel "Das Kranzbach"), 2021.
[17]    Darlichow, I. & Booth, M., Aura-Soma, Heilung durch Farbe, Pflanzen-, und Edelsteinenergie (Aura-Soma, healing through colour, plant and crystal energy), Knaur, München, 2001.
[18]    Aura-Soma, Homepage. https://www.aura-soma.com/en. Accessed on: 27 Mar. 2022.
[19]    Shinohara, S., About the Aura-Soma Academy, Dev Aura, and the Shire Farm. Answering the author's questions via email reply on 15 and 23 Mar. 2022 (Personal Assistant to the Chairman of Aura-Soma Products Ltd), 2022.

# IMPORTANCE OF SOCIAL INTERACTION AND INTERCULTURAL COMMUNICATION IN TOURISM

MAURO DUJMOVIĆ & ALJOŠA VITASOVIĆ
Faculty of Economics and Tourism, Juraj Dobrila University of Pula, Croatia

## ABSTRACT
Tourism as a social phenomenon may be defined as a special form of indirect contact between two societies which may be comprehended according to the following scheme: communicator–message–receiver. Therefore, the primary focus of tourism is the communication between tourists and the host destination. Tourism is a result of the interaction of people from emissive countries with the receptive countries. Establishment of contacts with people from different countries and cultures in contrast to the anonymity and alienation encountered in everyday life turns out to be a very important motivation for travelling. The scope of this article is to highlight the importance of the social interaction and effective intercultural communication in tourism encounters. Although tourism encounters may result in mutual appreciation, understanding, respect, tolerance and the overall improvement of the social interactions between individuals, it also represents a potential minefield full of difficulties, which occur mainly due to cultural differences in communication and rules of social behaviour.
*Keywords: intercultural, communication, interaction, cultural differences, host, guest, contact, culture.*

## 1 INTRODUCTION

Tourism is primarily a social activity including a temporary migration of people to places outside their everyday surroundings and including pleasure deriving from the participation in various leisure activities, facilities and services provided to cater to tourists' needs. The study of tourism is the study of people away from their usual surroundings, the study of establishments which have been set up in response to the needs of tourists and of the impacts that they have on the economic, environmental and socio-cultural wellbeing of host destinations [1].

Until recently, participation in tourism was limited to the chosen few who could afford time and money to travel to other destinations. Contemporary tourism has evolved as a consequence of various benefits enjoyed by laborers as a part of social welfare policies, that is, people's right to take leave for paid annual vacation. The development of the means of transportation, the development of accommodation facilities and the growth of package tour holidays have all contributed to the rise of tourism, that is the travel for pleasure. People's choices of travel and holidays mainly depend on their standard of living, their profession, the level of their education and their disposable income. Today every seventh person in the world is a tourist. Tourism has ceased to be the privilege of a few and has become one of the most popular activities of people in their leisure time.

Tourism is of major economic and social importance. With a world growth rate in international visitors' arrivals of approximately 5% per annum, tourism has become one of the fastest growing economic activities globally and it has proven to be resilient to political and natural crises and disasters, recovering quite rapidly once these calamities have passed [2]. It is the most important export industry and earner of foreign exchange in many countries all over the world. Tourism has triggered employment, investment and entrepreneurial activity, improved the economic structure and made positive contribution to the balance of payment in many countries throughout the world. However, the unprecedented growth of tourism has given rise to a multitude of economic, environmental and socio-cultural impacts which are concentrated in destination areas. In the past, tourism was encouraged for its

WIT Transactions on Ecology and the Environment, Vol 256, © 2022 WIT Press
www.witpress.com, ISSN 1743-3541 (on-line)
doi:10.2495/ST220071

economic benefits with little consideration for the consequences on host communities and their environments. For this article's sake the economic and environmental consequences on the host destination shall be purposefully put aside and sociocultural effects brought to the fore.

Sociocultural impacts appear as a result of particular types of social relationships that occur between tourists and hosts as a result of their mutual contact. In this respect tourism may be defined as a special form of indirect contact between two societies which may be comprehended according to the following scheme: communicator–message-receiver [3]. Tourism is based on the interaction and communication of people from emissive countries with people from receptive countries. This interaction offers new possibilities of encountering various people and encouraging the exchange of values and experiences between individuals originating from different cultural backgrounds. Tourism brings people together, making them more tolerant and more open for other cultures and in this way, it has become the promotor not only of worldwide development but worldwide peace as well [4].

The scope of this article is to highlight the importance of the social interaction and the effective intercultural communication in tourism encounters. Although tourism encounters may result in mutual appreciation, understanding, respect, tolerance and the overall improvement of the social interactions between individuals, it also represents a potential minefield full of difficulties, which occur mainly due to cultural differences in communication and rules of social behaviour.

## 2 TOURIST–HOST ENCOUNTERS

According to Wall and Mathieson [5] the most relevant categories in the research on the social and cultural impacts of tourism include: the tourist and their motivations, attitudes and expectations as well as their purchasing decisions; the host, that is the inhabitants of a destination often employed in tourism industry and their provision of labour services and the tourist–host relationship focussing on the nature and contacts between hosts and guests and the consequences of these contacts.

The focus of this article is on the latter two topics and it aims to highlight the importance of the social interaction and effective intercultural communication in tourism encounters contributing to the achievement of the goals set out by World Tourism Organisation in Manilla Declaration that tourism "stands out as a positive and ever-present factor in promoting mutual knowledge and understanding and as a basis for reaching a greater level of respect and confidence among all the peoples of the world" [6].

The most common social and cultural consequences of the tourism development are reflected in the changes of the quality of life of a local destination and its values, norms and traditions, individual behaviours and lifestyles, traditional ceremonies, community organisation and social interaction. Sociocultural impacts spring up from the social interaction and contact between tourist and hosts, which occur in three main contexts: during the purchase of a good or service from the host, when tourists and hosts coexist side by side, for example, on a beach or at a pub and when two parties come face to face with the aim of exchanging information and ideas [7]. The first two kinds of contact are more common and related to mass tourism and by character rather superficial and short. The third type of contact requires more effort and a deeper involvement with the other party and it is normally regarded as indispensable tool for increasing international understanding in tourism [8].

Despite the fact that the analysis of tourist–host encounters is a difficult task, it is possible to make some general observations about the nature of these encounters. It is logical to assume that the greater the economic, cultural and social differences between a tourist and a member of the destination community, the less balanced will be the relationship between them. In most tourist–host encounters the tourist is on vacation, easy-going, enjoying their leisure and the experience of being in a different place, whereas the host is relatively stationary and often employed in the tourism industry in which case they spend a significant amount of time catering to the needs and desires of tourists [9]. Under such conditions, the contact is transitory in nature, it suffers from temporal and spatial constraints, it lacks spontaneity and the relationship is unequal and unbalanced, providing little opportunity for deeper interaction [10]. Furthermore, tourism is usually constrained by certain season of the year limiting tourists' length of stay. Tourist–host contact may also be limited by the location and span of tourist-related services often concentrated in a small number of complexes, which are commonly referred to as tourist zones or ghettos, where tourists are isolated and discouraged to mingle with the local population reducing the possibility of the two parties to come into deeper contact. Tourists tend to buy into pre-packaged tours and pre-planned attractions and thus invest into convenient, comfortable and risk-free tourist experiences, provided at the expense of less frequent and spontaneous contacts with their hosts. Relationships which were once motivated by traditional hospitality may become commercialised and reduced to a series of cash-generating activities. Material inequality between tourists and their hosts often results in a tendency for the tourist–host relationship to be unequal and unbalanced in character. That is, local people may feel intimidated and inferior when faced with tourists' obvious wealth and may feel offended or frustrated in comparison to tourists who are on holiday and even develop a hostile stance towards incoming tourists [11].

There are also other factors which promote or impede friendly host–guest relationship and contacts, such as the size of tourism development, types of tourists and tourism, the degree of tourists' ghettoization, the length of stay, residents' involvement in tourism development, language and communications, etc.

Social interaction includes communication among people on a daily basis. The main goal of social interaction is to come in contact with other people in various situations, participate in conversation, exchange attitudes and views, learn about each other's social and cultural background, develop relationships, etc. [12]. Successful social interaction may contribute to the eviction of social and national prejudices and the promotion of better intercultural understanding, social inclusion and positive social change. It is of great importance for the promotion and contribution of understanding between tourists and their hosts.

## 3 INTERCULTURAL SOCIAL INTERACTION AND COMMUNICATION IN TOURISM

It is impossible to be a tourist in isolation and it is inevitable that tourists come into contact with other people. As already mentioned these may be either other tourists or members of local communities or more likely it will be a combination of the former and the latter taking place in planes and buses, hotels and restaurants, tourist attractions, shopping centres or nightclubs. Hosts can be local inhabitants, investors, developers and those who are employed in tourism industry such as hoteliers, front desk employees, waiters, shop assistants, tour guides taxi and bus drivers, etc. Therefore, social interaction in tourism can occur between tourist and other tourists, tourists and service providers, tourists and local residents, tourists and workers in the tourism industry and tourists and investors [13].

The opportunities for successful mutual interaction depend on the personal traits of tourists and hosts alike, such as tolerance, enthusiasm, generosity, welcoming attitudes, willingness to listen, mutual respect and tolerance, whereas resentment, disrespect, lack of appreciation for each other's cultural background, arrogance, and sense of superiority diminish the chances for interaction. In addition, motivation plays a significant role in the achievement of a proper social interaction between tourists and hosts. In most cases tourists meet hosts with a lack of proper motivation to do it limiting themselves to unavoidable conversations with fellow tourists or most often with service providers in various tourist facilities or they may like to get involved in conversation but with no further commitments. As a rule, only a few tourists show the willingness to engage in deeper and longer conversations with hosts in order to get acquainted with each other, share personal experiences and develop long-term friendships. In this process social interaction is influenced by cultural values, which play a dominant role as well [14].

Participants' interests and their willingness to be involved in deeper and more demanding social interaction is determined by cultural values, which stimulate the development of particular attitudes towards people [15]. The so-called cultural distance [16], that is cultural similarity or cultural gap between guests and hosts is a very important factor which defines the intensity of their mutual interaction in tourism. The bigger the gap the more unbalanced and less intense interaction between guests and hosts is likely to emerge and vice versa the smaller the cultural gap the more balanced and more in-depth interaction is likely to develop.

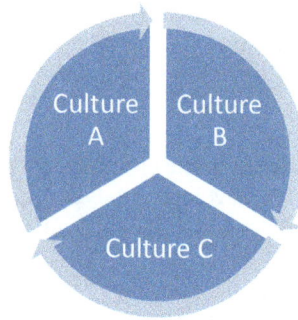

Figure 1:  The cross-cultural interaction model [12].

However, cultural disparity can also play an important role in attracting people to each other and result in dismantling of the social and cultural obstacles [17]. Such interaction involves individuals with different cultural backgrounds in regard to their values, communication styles, expectations, perceptions, rules of behaviour and the way they understand the world. They carry different cultural conventions and norms, speak different languages and use different symbols of communication. Their interaction is intercultural because every person involved in the interaction is different in regard to their cultural background. Therefore, the interaction is least intercultural if the involved individuals are culturally very similar to one another and most intercultural if the involved individuals are culturally significantly different from one another [18]. Such a face to face communication and interaction between participants with different cultural backgrounds may result in both positive and negative outcomes. Mutual affection, understanding, respect tolerance, development of positive attitudes and reduction of ethnic prejudices, stereotypes and racial tension, improvement of the social interactions between individuals, cultural enrichment and learning about others are most commonly singled out as positive outcomes contributing to the development of pride in the local culture on the part of the local residents and a possible development of personal relationships and long-term friendships [19]. On the other hand, in destinations (usually in less developed countries) where cultural differences between tourists and hosts are more pronounced then elsewhere the negative outcomes may be greater varying from exploitation, mistrust, stereotyping and unmeaningful conversation to lack of respect for local values and lifestyles on the part of tourists who are perceived as aggressive and insensitive by the locals [20].

Mass tourism, which is the dominant form of tourism in the world, offers only minor opportunities to learn about the social, cultural, and political conditions of the destinations which are visited. Instead of deconstructing various local residents' established misperceptions and stereotypes about incoming tourists, tourism perpetuates them and often creates new prejudices, stereotypes, misunderstandings, hostility, tensions and conflict of values and communication problems [21].

Language is an important tool of communication and it constitutes a part of the social and cultural attributes of any population. Therefore, proper intercultural communication between tourist and their hosts may contribute to the promotion of versatile changes in local culture while maintaining and preserving local ethnic and cultural identity [22]. That is the reason why proper intercultural communication turns out to be of great importance to the development and advocacy of tolerance and understanding between tourists and their hosts.

## 4 IMPROVING INTERCULTURAL COMMUNICATION IN TOURISM

In situations when tourists and hosts are conscious of their differences and when they try their best to understand them and each other, the difficulties in social interaction may be avoided or minimized. People should be more approachable and prone to accept new ideas and, in this way, make a personal contribution towards successful intercultural communication. Adaptability, tolerance, respect for other peoples' viewpoints are crucial qualities for all those involved in the process of social interaction and communication in tourism. As it was pointed out hereinbefore, there is no tourist in isolation from other people whom he comes into contact with and who are potentially different from him. In order for the communication to be effective and fair and if all the involved want to benefit from their mutual interaction their diverse backgrounds must be acknowledged and respected [23].

Tourists and hosts will be ready to acquire new ways of interacting outside the already established patterns of behaviour only if they disregard and avoid stereotyping, ethnocentrism and even racism and only if they bear in mind that they are confronted with an individual and culture different from their own. It will help them to broaden their horizons and open up to new opportunities and fresh perspectives. In such a way they will also more readily admit that their own attitudes and points of view are not necessarily always the best ones, which will make them more attentive to other peoples' necessities and make them restrain their mouths to work faster than their brains. They will learn to listen more attentively, to show mutual appreciation and to analyse each other more properly becoming in this way more effective intercultural communicators.

## 5 CONCLUSION

Tourism is a unique social and cultural event for the tourist and the host [24]. Tourism is a social process which brings together different people from all over the world in a form of social interaction. Although it is often regarded as a very important tool of economic growth and development, it is also an agent of social change and cultural change and a tool for the promotion of international peace and understanding. Naturally, tourism is rarely the only agent of social and cultural change. No society and culture are immune to progress and development in which process the rapid growth of population, dissemination of multinational corporations, urbanisation and industrialisation are the most important factors which have greater impact on the societies than tourism. The current conditions of globalisation, with pronounced mobility and ephemerality allow for a much greater degree of mixing and interchange than it was possible in the past. Mobility is a prerequisite of tourism and it is indispensable for the social contact and interaction between people of different cultural backgrounds.

Travels are characterised by interaction and communication between tourist and other tourists, tourists and service providers, tourists and local residents, tourists and workers in the tourism industry and tourists and investors. Good communication doesn't only consist of conveying information but also of mutual understanding of communication participants' wishes and needs. If one of the parties takes a dominant stand and disregards the intercultural dialogue, communication may be difficult and it may have significant negative impacts for the development of tourism in a destination leading to the situation in which local residents may perceive tourists as intruders in their own culture. People's own cultural backgrounds, their attitudes, values, norms and believes determine the way in which they behave in the process of communication. Situations in communication which include individuals of different cultural backgrounds are called intercultural communication situations [25].

Intercultural communication is a symbolic, interpretive, transactional, contextual process, in which people from different cultures create shared meanings [26]. It is never perfect and the difficulties in intercultural communication are caused by cultural differences in verbal and non-verbal signals, wrong interpretations, relationship patterns, communication gaps, conversation and interaction styles, the activation of negative stereotypes, cultural values and time and context situations. On the other hand, effective cross-cultural communication between tourist and their hosts may promote adaptive changes in local culture while preserving or revitalising local ethnic and cultural identity [27]. The achievement of this goal doesn't only require the acquisition of cultural knowledge or tolerance towards cultural differences. Successful intercultural communication requires from people to be accessible to new experiences and to be ready to accept their own and other's unique and different identities. In this way the interaction with other cultures may change peoples' innate perceptions and attitudes contributing to their own cultural and spiritual development. In conclusion it is therefore possible to point out that the proper intercultural communication is an inevitable prerequisite for the promotion of tolerance and understanding between tourists and their hosts in any destination.

## REFERENCES

[1]   Mason, P., *Tourism, Impacts, Planning and Management*, 4th ed., Routledge: London, 2020.
[2]   Sharpley, R., *Tourism, Tourists and Society*, Routledge: London and New York, 2018.
[3]   Jadrešić V., *Turizam u Interdisciplinarnoj Teoriji i Primjeni*, Školska Knjiga: Zagreb, 2001.
[4]   Urry, J. & Jonas, L., *The Tourist Gaze 3.0*, SAGE Publications: London, 2011.
[5]   Wall, G. & Mathieson, A., *Tourism: Change, Impacts and Opportunities*, Pearson Education: Harlow, 2006.
[6]   WTO, Manila Declaration on World Tourism, 1980. https://www.univeur.org/cuebc/downloads/PDF%20carte/65.%20Manila.PDF. Accessed on: 20 Feb. 2022.
[7]   Hannam, K. & Knox. D., *Understanding Tourism*, SAGE Publications: London, 2010.
[8]   Aramberri, J., The host should get lost: Paradigms in the tourism theory. *Annals of Tourism Research*, **28**(3), pp. 738–761, 2001.
[9]   Reisinger, Y. & Turner, L., *Cross-Cultural Behaviour in Tourism*, Routledge: Abingdon, 2012.

[10]   UNESCO, The effects of tourism on socio-cultural values. *Annals of Tourism Research*, **4**, pp. 74–105, 1976.
[11]   Reisinger, Y., *The Host Gaze in Global Tourism*, CABI: Wallingford, 2013
[12]   Reisinger, Y., *International Tourism, Cultures and Behaviours*, Elsevier: Oxford, 2009.
[13]   De Kadt, E., *Tourism: Passport to Development?* Oxford University Press: New York, 1979.
[14]   Zain Sulaiman, M. & Wilson, R., *Translation and Tourism*, Springer Nature: Singapore, 2019.
[15]   Berno, T., When a guest is a guest. *Annals of Tourism Research*, **26**(3), pp. 656–675, 1999.
[16]   Smith, M., Macleod, N. & Hart Robertson, M., *Key Concepts in Tourist Studies*, SAGE Publications: London, 2010.
[17]   OECD, *The Impact of Culture on Tourism*, OECD Publishing: Paris, 2009.
[18]   Lustig, M.W. & Koester, J., *Intercultural Competence: Interpersonal Communication Across Cultures*, 5th ed., Shanghai Foreign Language Education Press: Shanghai, 2007.
[19]   Cohen, E., Towards a sociology of international tourism. *Social Research*, **39**(1), pp. 164–182, 1972.
[20]   Pearce, P.L., The relationship between residents and tourists: The research literature and management direction. *Global Tourism*, ed. W.F. Theoblad, Butterworth-Heinemann: Oxford, pp. 110–124, 1988.
[21]   Dodds, R. & Butler, R., *Overtourism*, De Gruyter: Berlin, 2019.
[22]   Evans, N., Tourism and cross cultural communication. *Annals of Tourism Research*, **3**, pp. 189–199, 1976.
[23]   Lohmann, G. & Panosso Netto, A., *Tourism Theory, Concepts, Models and Systems*, CABI: Wallingford, 2017.
[24]   Murphy, P., *Tourism: A Community Approach*, Routledge: London, 1985.
[25]   Marinov, V., Vodenska, M., Assenova, M. & Domagradjieva, E., *Traditions and Innovations in Contemporary Tourism*, Cambridge Scholars Publishing: Newcastle upon Tyne, 2018.
[26]   Albu, C.E., International communication in tourism. *Cross Cultural Management Journal*, **XVII**(I), pp. 7–14, 2015.
[27]   Shahzalal, Md., Positive and negative impacts of tourism on culture: A critical review of examples from the contemporary literature. *Journal of Tourism, Hospitality and Sports*, **20**, pp. 30–34, 2016.

# STRATEGIC MANAGEMENT OF TOURISM SUSTAINABILITY THROUGH THE GREEK STAKEHOLDERS' PERSPECTIVE ON THE IMPACTS OF EVENTS: THE CASE OF PATRAS' CARNIVAL, GREECE

EVANGELIA PAPPA[1], ELENI DIDASKALOU[2], GEORGIOS KONTOGEORGIS[3] & IOANNIS FILOS[1]
[1]Department of Public Administration, Panteion University of Social and Political Sciences, Greece
[2]Department of Business Administration, University of Piraeus, Greece
[3]Department of Tourism, Ionian University, Greece

## ABSTRACT

Events can play a critical role in implementing sustainable developing models at destinations. Furthermore, the perceptions of stakeholders may contribute to the sustainability of a tourist destination in the long term. This paper presents an insight into the stakeholders' perceptions of the importance of festival events in promoting tourism sustainability, concentrating on Patras' Carnival. Festivals can be an instrument for tourism development, city image improvement and boosting regional economies. Based on a theoretical model that is grounded in the social exchange theory the research enriches the existing knowledge on promoting sustainable events and sustainable approaches to tourism development by taking into account the views of the leading players of tourism, local residents and business owners. A quantitative survey via a structured questionnaire was conducted in the city of Patras, Greece's third-largest city before the COVID-19 pandemic outbreak. The questionnaire was distributed to the residents and business owners during the Patras' Carnival (Patrino Carnavali), the largest event of its kind in Greece. In total, 238 people participated in the study. It will be presented as a variety of positive and negative impacts of tourism toward economy, society, culture and environment and shed light on adequate managerial practices that boost further tourism flows in cities. The results may be useful not only to local government entities involved in the tourism strategic planning but also to stakeholders engaged in creating sustainable competitive advantage in the tourism industry.
*Keywords: strategic management, sustainability, tourism industry, stakeholders' perspective.*

## 1 INTRODUCTION

Nowadays, festival events enhance residents' "community satisfaction", through regional and neighbourhood development promoting the cultural heritage, of the people of the area, in various forms [1]. Festivals and carnivals are classified as a type of cultural tourism [2].

In this context satisfaction for locals' arise maximizing the benefits and minimizing the costs of tourism. Taking into account the social exchange theory one can better understand positive and negative outcomes of tourism toward economy, society, quality-of-life and environment, so as to develop a long-benefit for the destination [3], [4].

The focus of the study is Patras' Carnival, which is a great cultural event, not only for Patras city, but also for the region of western Greece. The purpose of the study is to assess the perceptions of locals' about carnivals, towards sustainable tourism development. Thus, analysing resident's impact perceptions of tourism development that contribute to improving the image of the festival event as a tourist destination. Findings may help destinations on promoting sustainable events and planning for additional development.

WIT Transactions on Ecology and the Environment, Vol 256, © 2022 WIT Press
www.witpress.com, ISSN 1743-3541 (on-line)
doi:10.2495/ST220081

## 2 LITERATURE REVIEW

### 2.1 Cultural tourism and carnivals

According to Britannica the historical origin of carnival is not exactly known [5]. In Greece the carnival has its roots in ancient times and is inextricably linked to the worship of the god Dionysus. The English word "carnival" comes from the Latin "carnem levare" or "carnis levamen", which means an abstention from meat. In Greek, the word "carnival" has the same meaning and is connected with the beginning of Lent. The largest carnival in Greece is Patras' Carnival [6]. Masquerades and parades are common features of carnivals all over the world [7]. Carnival is a form of tourism product and should be examined in the context of special event tourism [8], [9]. Special event tourism has many impacts not only on the economy of a destination [10] but furthermore, for the communities in which they take place, they can play a vital role for the sustainable development of the destinations [11], [12]. The implications of special events according to Shone and Parry [13] are given in Fig. 1

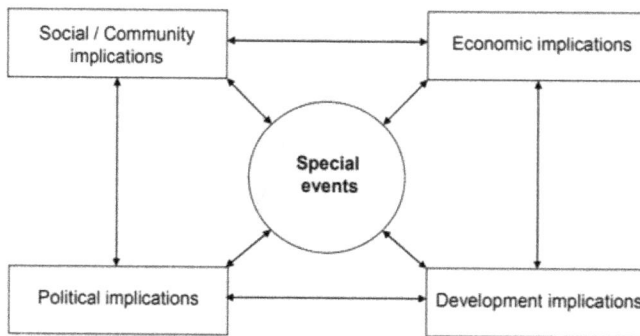

Figure 1:   The implications of special events. *(Source: Adapted from Macgregor et al. [14].)*

Carnivals, as an urban festival, have many impacts on various sectors of a city (educational, artistic, social, political, economic) [15], [16]. Furthermore, there are positive and negative impacts which generally can be clustered into socio-cultural, political, economic, physical categories [7]. Also, carnivals can differentiate and promote a specific identity of a region and enhance town's heritage.

### 2.2 Impacts of carnivals

As the topic of sustainability, nowadays, is an important issue, culture is a key point for urban development, making cities creative, attractive and sustainable [17]. Festival tourism as a form of cultural tourism plays significant role in achieving regional economic development and community, thereby enhancing place identity and local attractiveness [18]. However, it should not be ignored the fact that festivals, carnivals and events are significant emitters of greenhouse gases, due to transportation and energy use [19]. But, it has also to be mentioned that there are examples of environmental improvements, e.g. new infrastructure and restoring historic buildings [20]. On promoting sustainable events is crucial to understand stakeholders' perspective on the impacts of the events. Furthermore, for taking a range of measures to boost festival/carnival tourism is important to deepening the knowledge about these events and their impacts [21], [22]. Impacts are usually classified into three categories: environmental, economic and sociocultural.

To estimate and quantify environmental impacts of events various approaches can be used such as environmental impact assessment (EIA), and life cycle analysis (LCA), biophysical methods and carbon emissions. As so, ecological footprint and carbon footprint are tools that consider environmental impacts that occur beyond a festival site [23].

Benefits of festivals include the generation of income, creation of jobs and also the enhancing of communal cohesion. However may attract people who behaves badly or in a way that breaks the law, causes cultural adulteration, worsen the city's sanitation through pollution, vehicular and human congestion and influence an increase in the cost of living [24].

### 2.3 Social exchange theory and tourism

The theory of social exchange is one of the oldest theories of social behaviour which also incorporates the concept of interaction. In an exchange process actors are dependent on each other for outcomes they value. They act in a way that increases positive outcomes and decreases negative outcomes. In terms of tourism, there are studies which claim that the meeting between the host community and visitors can either be an opportunity for rewards and satisfying exchanges, or it can be a means of boosting the local community's impulse to exploit visitors [25], [26]. Furthermore, in the context of social exchange theory, economic, social, and environmental impacts affect the perceptions of locals for tourism. A theoretical model was proposed by Yoon et al. (Fig. 2) which examines the relationship among the aspects of perceived tourism impacts, total impacts, and support for tourism development [27].

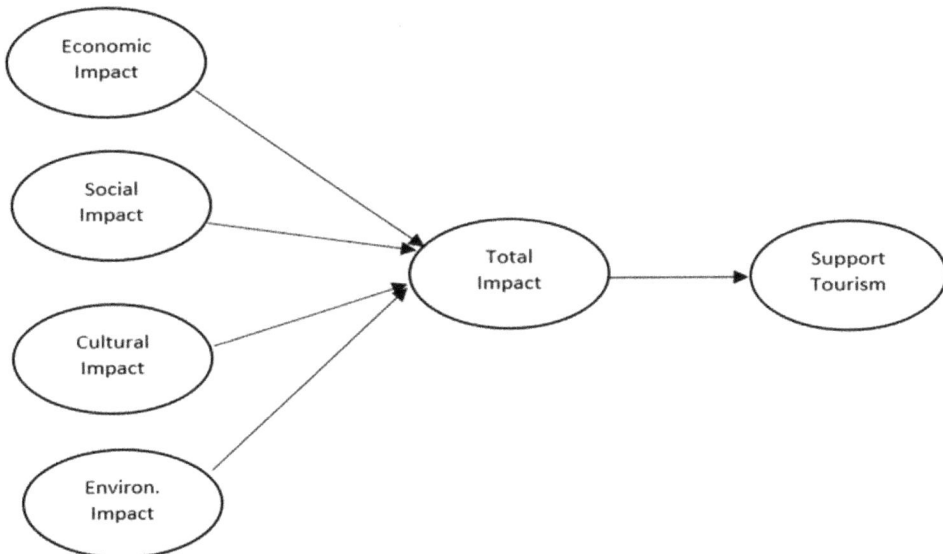

Figure 2:   Theoretical model for perceptions of residents towards tourism development. *(Source: Adapted from Yoon et al. [27].)*

Facts that influence residents' attitudes towards tourism are [28]:

- Community attachment or length of residence;
- Knowledge about tourism, contact with tourists and concentration of visitors;
- Personal reliance on tourism;
- Proximity with the tourism centre;
- Level of participation in recreational activities;
- Demographic variables;
- Seasonality;
- Tourism taxes and perceived future of the community;
- The development stage of a destination.

## 3 METHOD AND MATERIAL: STUDY AREA

### 3.1 Study area: Patras' Carnival

Patras is the capital of the Prefecture of Achaia and is the third largest city in Greece. The population of Achaia is about 297,000 people and has an area of about 3.274 km². It has 7,345 hotel beds and 1,546 beds for rent. Analytically, it has 19 five stars hotel beds, 2,691 four stars hotel beds, 2,584 three stars hotel beds, 1,899 two stars hotel beds and 152 one star hotel beds. It has 449 tourist furnished houses (villas) and 404 camping places too. The total of overnight stays in hotel accommodation (foreigners) is 250,065 and the total overnight stays in hotel accommodation (locals) are 65,455. Finally, it has two seaports (the port of Patras and the port of Aegio) and one airport in Araxos [29].

Patras has a developed tourist infrastructure, but it is not overdeveloped. The tourism of Patras is based mainly on Greek visitors. The airport of Patras has very low passenger traffic and the tourist traffic of Patras is based on the Greek tourist market. A significant percentage of the tourist traffic of Patras comes from the carnival festival that it has been organizing successfully for many years.

Patras' Carnival is an event that has a 98.4% recognition nationwide, counts about 100,000 visitors, of which 50.0% comes from Athens, 40.0% are young people and students. During the carnival of 2009, visitors spent about 17 million Euros, while crew members approximately 2.5 million Euros and the Municipality of Patras spent about 1.2 million Euros. In addition 250 jobs was created [30].

The carnival of Patras is a great cultural event with about 120,000 participants and takes place every year [31]. The beginning of the Patras carnival dates back to 1829, at that time the merchant named Moreti was organizing carnival dances at his house in Patras. The union of the Ionian Islands with Greece in 1864 and the port of Patras resulted in the settlement of many westerners in Patras. Many of them were from Italy which had a long tradition in organizing carnival events and so they transferred the cultural tradition of the carnivals to Patras [32].

The carnival of Patras is considered one of the largest in Europe and stands out for the variety of events that a family can participate in [33]. In other words, it includes parades of children and adults, the hunt for lost treasure, many parties in the streets of the city, parade of dance schools and visitors have the opportunity to participate in the events [31]. Visitors can participate in the carnival by joining in the carnival parade, taking part in parties and joining in various events that take place in the city of Patras. The participation of the guests is booked with the payment of a fee. In addition, the Patras' Carnival includes events such as the children's carnival, street theatres, concerts, theatres and other events [32]. All the above highlight the uniqueness of the city of Patras as a carnival tourist destination and the opportunities and possibilities it has.

## 3.2  Data and instrument

The target population of the study refers to Patrician residents and business owners of different tourism activities. A face-to-face survey was conducted during the carnival festival period, namely from January to February 2018. A total of 279 responses were gathered, out of which 238 were usable. The basic research instrument for data collection was a structured questionnaire. It was developed following the multiphase mixed approach of Churchill [34], [35] and DeVellis and Thorpe [36]. This method involves a systematic process of formulating a valid and reliable research instrument. In essence, four successive stages are applied in the questionnaire development: (i) extensive literature review, (ii) items generation and scales construction, (iii) preliminary pilot testing and (iv) distribution to an actual sample.

The questionnaire items were originally borrowed from the English literature and were translated by the authors to Greek based on Brislin's guidelines [37], [38]. First, a bilingual language expert reviewed and revised the translation to guarantee that the translated version reproduced the clarity, comprehensibility and suitability of the original items. Furthermore, the appropriateness of the measures for each conceptual variable was confirmed by content experts (tourism professors). The final version of the questionnaire eliminated any possible mistakes, ambiguities based on the recommendations of the experts [39]. The questionnaire administration took approximately 10 minutes to complete.

The questionnaire consists of 44 items organized in three blocks: classificatory data about participants (7); perception of tourism impacts (36); and the overall attitude towards tourism development (1). Attribute items were extracted from extant studies of residents' attitude towards tourism impacts. Perception and attitude have been measured using five-point Likert scale, as recommended by Maddox [40] and Williams and Roggenbuck [41], anchored by 1 (strongly disagree) to 5 (strongly agree).

## 4  RESULTS

The survey data have been analyzed with IBM SPSS Statistics 25.0 applying both descriptive and inferential statistic techniques. First, descriptive statistics were performed to analyse respondents' socio-demographic profile. Second, the 36 attitude items were factor-analysed to identify the perceived impacts of tourism by Patrician residents and bossiness owners. In Table 1 we can observe demographic characteristics of the participants in the survey such as gender, age, education, marital status and occupation.

The underlying factors of perceived tourism impacts of residents in Patras were examined by the exploratory factor analysis (EFA). The Kaiser–Meyer–Olkin (KMO) measure of sampling adequacy and the Bartlett's Test of Sphericity were included in the analysis to determine the fitness of the data [42]. The KMO stands on 0.882 and Bartlett's test of sphericity was significant (p value <.000), indicating that the correlations between items are sufficiently large [43]. A principal component analysis with varimax rotation method was applied to confirm the scale construct validity. The EFA revealed two factors with the total variance explained of 67.023%, suggesting a satisfactory factor solution.

Table 2 presents the derived two factors with their loadings and also the means and standard deviations of the 36 items used. Each factor was characterized according to the core items constituting it [43]. The first factor named "positive impacts" is related to

Table 1: Respondent's profile.

| Variable | Category | N | Percentage |
|---|---|---|---|
| **Gender** | Male | 114 | 47.90 |
| | Female | 124 | 52.10 |
| **Age** | Up to 24 | 44 | 18.50 |
| | 25–34 | 85 | 35.70 |
| | 35–44 | 64 | 26.90 |
| | 45–54 | 30 | 12.60 |
| | Above 54 | 15 | 6.30 |
| **Education** | High School | 94 | 39.50 |
| | University | 85 | 41.20 |
| | Postgraduate | 64 | 9.70 |
| | PhD | 30 | 2.10 |
| | Other | 15 | 7.50 |
| **Marital status** | Unmarried | 137 | 57.60 |
| | Married | 35 | 14.70 |
| | Married with children | 56 | 23.40 |
| | Divorced | 10 | 4.20 |
| **Occupation** | Tourism oriented | 121 | 50.80 |
| | No tourism oriented | 117 | 49.20 |
| **Tourism-oriented sector** | Accommodation Services | 20 | 16.50 |
| | Transport services | 8 | 6.60 |
| | Cultural Activities | 6 | 5.00 |
| | Athletic Activities | 9 | 7.40 |
| | Travel planning services | 3 | 2.50 |
| | Food Services | 40 | 33.10 |
| | Conferences and exhibitions | 1 | 0.80 |
| | Retail supply of goods | 28 | 23.10 |
| | Other | 6 | 5.00 |

positive aspects including employment opportunities, economic growth of the region, environmental protection and maintenance of cultural heritage. The second factor is "negative impacts", which focuses on negative aspects of festival event including drug trafficking, crime problems, labour exploitation, parking problems, increase in property rents and price of goods and services.

## 5 FINAL REFLECTIONS AND RECOMMENDATIONS
The study depicts local perception of impact attitudes towards Patras' tourist destination. The number of cultural events, the preservation of cultural identity and heritage and also the local attractiveness result in boosting Patras' image. Furthermore, the investment in tourism

infrastructure, the increasing of job offered stimulate the place branding, making Patras not only a destination to visit, but also to live and work.

Nevertheless, stakeholders such as the Municipality of Patras, the Region of Western Greece, the tourist and professional organizations should all work together as to eliminate the negative effects of tourism and increase locals' satisfaction, Also, they should take drastic measures to avoid damaging natural surroundings and landscape. Finally, the responsible public authorities must control the phenomenon of exorbitant prices and impose the appropriate punishment on the perpetrators.

Table 2:  Results of the exploratory factor analysis.

| Measure items<br>*Tourism at Patras causes….* | | Loadings | Mean | SD |
|---|---|---|---|---|
| **Factor 1: Positive impacts (PO)**<br>*(EigenValue: 15.141; Total variance: 29.369)* | | **0.717** | **2.767** | **1.170** |
| *Economic* | **POEC1.**increase in employment opportunities | 0.890 | 2.735 | 1.296 |
| | **POEC2.**improvement of the economic status | 0.859 | 2.765 | 1.237 |
| | **POEC3.**improvement of local government resources | 0.858 | 2.671 | 1.240 |
| | **POEC4.**attraction of investments | 0.852 | 2.609 | 1.234 |
| | **POEC5.**improvement of the quality of life | 0.727 | 2.559 | 1.195 |
| | **POEC6.**improvement of residents' purchasing power | 0.605 | 2.630 | 1.161 |
| *Social* | **POSO1.**improvement of the quality of public services | 0.695 | 2.345 | 1.086 |
| | **POSO2.**improvement of public infrastructures | 0.682 | 2.786 | 1.144 |
| | **POSO3.**improvement of the road and transport infrastructure | 0.667 | 2.979 | 1.171 |
| | **POSO4.**increase of choices within the local market | 0.634 | 2.782 | 1.130 |
| | **POSO5.**the influence inhabitants' professional orientation to tourism-related jobs | 0.615 | 2.878 | 1.143 |
| | **POSO6.**improvement of residents' professional skills | 0.572 | 2.752 | 1.148 |
| | **POSO7.**increase of occasions of amusement | 0.526 | 3.000 | 0.994 |
| *Cultural* | **POCU1.**the construction of modern buildings | 0.817 | 2.418 | 1.252 |
| | **POCU2.**conservation and restoration of historical monuments | 0.808 | 2.941 | 1.189 |
| | **POCU3.**conservation and restoration of historical places and monuments | 0.803 | 2.866 | 1.225 |
| | **POCU4.**maintenance of cultural heritage | 0.765 | 2.975 | 1.176 |
| | **POCU5.**maintenance of the cultural identity | 0.728 | 2.983 | 1.132 |
| *Environmental* | **POEN1.**improvement of the general image of environment | 0.653 | 2.752 | 1.122 |
| | **POEN2.**protection of the environment | 0.639 | 2.681 | 1.143 |

Table 2: Continued.

| Measure items<br>*Tourism at Patras causes....* | | Loadings | Mean | SD |
|---|---|---|---|---|
| **Factor 1: Negative impacts (NE)**<br>*(EigenValue: 6.7093; Total variance: 13.013)* | | **0.524** | **3.132** | **1.180** |
| Economic | NEEC1.increase of property rents | 0.454 | 3.277 | 1.211 |
| | NEEC2.increase of the price of goods and services | 0.452 | 3.109 | 1.153 |
| | NEEC3.increase of the cost of real estate | 0.388 | 3.298 | 1.204 |
| | NEEC4.more spending to the region | 0.271 | 3.160 | 1.169 |
| Social | NESO1.negative effects in the ethos | 0.674 | 3.311 | 1.234 |
| | NESO2.negative influence in the lifestyle of locals | 0.626 | 3.403 | 1.168 |
| | NESO3.increase of drug trafficking | 0.615 | 3.034 | 1.259 |
| | NESO4.increase of road accidents | 0.577 | 3.252 | 1.130 |
| | NESO5.increase of criminality | 0.568 | 3.164 | 1.174 |
| | NESO6.increase of alcoholism | 0.541 | 2.966 | 1.158 |
| | NESO7.increase of labour exploitation of local population | 0.497 | 3.071 | 1.218 |
| | NESO8.increase of gambling | 0.468 | 3.445 | 1.160 |
| Environment | NEEN1.damage to the natural surroundings and the landscape | 0.661 | 3.382 | 1.201 |
| | NEEN2.increase of the pollution of the environment (rubbish, air pollution and noise) | 0.624 | 3.147 | 1.269 |
| | NEEN3.traffic congestion and parking problems | 0.523 | 3.172 | 1.162 |
| Culture | NECU1.negative effect in the cultural heritage | 0.445 | 3.618 | 1.180 |

Notes: KMO = 0.882; Bartlett's Test of Sphericity = 0.000; Overall variance explained = 67,023%. Extraction Method: Principal Component Analysis; Rotation Method: Varimax with Kaiser Normalization; * p< .05.

## ACKNOWLEDGEMENT
This work has been partly supported by the University of Piraeus Research Center.

## REFERENCES
[1]   Almeida, A., Teixeira, S.J. & Franco, M., Uncovering the factors impacting visitor's satisfaction: Evidence from a portfolio of events. *International Journal of Event and Festival Management*, **10**, pp. 217–247, 2019. DOI: 10.1108/IJEFM-01-2019-0002.

[2]   Esu, B. & Arrey, V., Tourists' satisfaction with cultural tourism festival: A case study of Calabar Carnival Festival, Nigeria. *International Journal of Business and Management*, **4**, p. 116, 2009. DOI: 10.5539/ijbm.v4n3p116.

[3]   Thibaut, J.W. & Kelley, H.H., *The Social Psychology of Groups*, John Wiley: Oxford, 1959.

[4]   Ziakas, V., Understanding an event portfolio: The uncovering of interrelationships, synergies, and leveraging opportunities. *Journal of Policy Research in Tourism, Leisure and Events*, **2**, pp. 144–164, 2010. DOI: 10.1080/19407963.2010.482274.

[5]   Britannica, Carnival: Definition, festival, traditions, countries and facts. https://www.britannica.com/topic/Carnival-pre-Lent-festival. Accessed on: 24 Jan. 2022.

[6]   VisitGreece, Carnival! The opportunity to have fun! https://www.visitgreece.gr/blog/travel-tips/601/carnival-the-opportunity-to-have-fun. Accessed on: 24 Jan. 2022.

[7]   Arcodia, C. & Whitford, M., Festival attendance and the development of social capital. *Journal of Convention and Event Tourism*, **8**, pp. 1–18, 2006. DOI: 10.1300/J452v08n02_01.

[8]   Getz, D., Special events: Defining the product. *Tourism Management*, **10**, pp. 125–137, 1989. DOI: 10.1016/0261-5177(89)90053-8.

[9]   Getz, D., Event tourism: Definition, evolution, and research. *Tourism Management*, **29**, pp. 403–428, 2008. DOI: 10.1016/j.tourman.2007.07.017.

[10]  Dwyer, L., Forsyth, P. & Spurr, R., Estimating the impacts of special events on an economy. *Journal of Travel Research*, **43**, pp. 351–359, 2005. DOI: 10.1177/0047287505274648.

[11]  Wallstam, M., Ioannides, D. & Pettersson, R.. Evaluating the social impacts of events: In search of unified indicators for effective policymaking. *Journal of Policy Research in Tourism, Leisure and Events*, **12**, pp. 122–141, 2020. DOI: 10.1080/19407963.2018.1515214.

[12]  Bowdin, G., Allen, J., Harris, R., McDonnell, I. & O'Toole, W., *Events Management*, 3rd ed., Routledge, 2012.

[13]  Shone, A. & Parry, B., *Successful Event Management: A Practical Handbook*, 3rd ed., Cengage Learning EMEA: Andover, UK, 2010.

[14]  Macgregor, C., Jones, R., Pilgrim, A. & Thompson, G., Assessing the environmental impacts of special events: Examination of nine special events in Western Australia, 2008. DOI: 10.13140/RG.2.1.3343.0004.

[15]  Kostopoulou, S., Vagionis, N. & Kourkouridis, D., Cultural festivals and regional economic development: Perceptions of key interest groups. *Quantitative Methods in Tourism Economics*, eds Á. Matias, P. Nijkamp & M. Sarmento, Physica-Verlag HD: Heidelberg, pp. 175–194, 2013. DOI: 10.1007/978-3-7908-2879-5_10.

[16]  Konsola, D. & Karachalis, N., Arts festivals and urban cultural policies in medium sized and small cities of Greece, 2009.

[17]  UNESCO, Culture: Urban future. Global report on culture for sustainable urban development, 2016.

[18]  Chi, X., Cai, G. & Han, H., Festival travellers' pro-social and protective behaviours against COVID-19 in the time of pandemic. *Current Issues in Tourism*, **24**, pp. 3256–3270, 2021. DOI: 10.1080/13683500.2021.1908968.

[19]  Duffy, M. & Mair, J., Future trajectories of festival research. *Tourist Studies*, **21**, pp. 9–23, 2021. DOI: 10.1177/1468797621992933.

[20]  Stankova, M.Z. & Vassenska, I., Raising cultural awareness of local traditions through festival tourism, 2015.

[21]  Amorim, D., Jiménez-Caballero, J.L. & Almeida, P., The impact of performing arts festivals on tourism development: Analysis of participants' motivation, quality, satisfaction and loyalty. *Tourism and Management Studies*, **16**, pp. 45–57, 2020.

[22]  Gu, X. et al., Evaluating residents' perceptions of nature-based tourism with a factor-cluster approach. *Sustainability*, **13**, p. 199, 2021. DOI: 10.3390/su13010199.

[23] Collins, A. & Cooper, C., Measuring and managing the environmental impact of festivals: The contribution of the ecological footprint. *Journal of Sustainable Tourism*, **25**, pp. 148–162, 2017. DOI: 10.1080/09669582.2016.1189922.

[24] Doe, F., Preko, A., Akroful, H. & Okai-Anderson, E.K., Festival tourism and socioeconomic development: case of Kwahu traditional areas of Ghana. *International Hospitality Review*, 2021. DOI: 10.1108/IHR-09-2020-0060.

[25] Nunkoo, R.. Toward a more comprehensive use of social exchange theory to study residents' attitudes to tourism. *Procedia Economics and Finance*, **39**, pp. 588–596, 2016. DOI: 10.1016/S2212-5671(16)30303-3.

[26] Sutton, W.A., Travel and understanding: Notes on the social structure of touring. International *Journal of Comparative Sociology*, **8**, pp. 218–223, 1967. DOI: 10.1163/156854267X00169.

[27] Yoon, Y., Gursoy, D. & Chen, J.S., Validating a tourism development theory with structural equation modeling. *Tourism Management*, **22**, pp. 363–372, 2001. DOI: 10.1016/S0261-5177(00)00062-5.

[28] Gabriel Brida, J., Osti, L. & Faccioli, M., Residents' perception and attitudes towards tourism impacts: A case study of the small rural community of Folgaria (Trentino – Italy). *Benchmarking: An International Journal*, **18**, pp. 359–385, 2011. DOI: 10.1108/14635771111137769.

[29] INSETE, Region of Western Greece: Annual report on competitiveness and structural adjustment in the tourism sector, 2021. (In Greek.)

[30] CityBranding.gr, Research on the financial benefits of "Patrino Carnival" 2012. https://www.citybranding.gr/2012/02/blog-post_20.html. Accessed on: 28 Jan. 2022.

[31] Papadimitriou, D., Service quality components as antecedents of satisfaction and behavioral intentions: The case of a Greek carnival festival. *Journal of Convention and Event Tourism*, **14**, pp. 42–64, 2013. DOI: 10.1080/15470148.2012.755885.

[32] CarnivalpatrasGr, Public Benefit Corporation. Patras Carnival, 2022. https://www.carnivalpatras.gr. Accessed on: 10 Feb. 2022.

[33] Carnivaland, Patras Carnival 2022, 2021. https://www.carnivaland.net/patras-carnival/. Accessed on: 10 Feb. 2022.

[34] Churchill, G.A., *Marketing Research Methodological Foundations*, 8th ed., Fort Worth, 2002.

[35] Churchill, G.A., A paradigm for developing better measures of marketing constructs. *Journal of Marketing Research*, **16**, pp. 64–73, 1979. DOI: 10.2307/3150876.

[36] DeVellis, R.F. & Thorpe, C.T., *Scale Development: Theory and Applications*, 5th ed., SAGE Publications: Thousand Oaks, CA, 2021.

[37] Brislin, R.W., Back-translation for cross-cultural research. *Journal of Cross-Cultural Psychology*, **1**, pp. 185–216, 1970. DOI: 10.1177/135910457000100301.

[38] Brislin, R.W., Comparative research methodology: Cross-cultural studies. *International Journal of Psychology*, **11**, pp. 215–229, 1976. DOI: 10.1080/00207597608247359.

[39] Peters, M., Chan, C.-S. & Legerer, A., Local perception of impact-attitudes-actions towards tourism development in the Urlaubsregion Murtal in Austria. *Sustainability*, **10**, p. 2360, 2018. DOI: 10.3390/su10072360.

[40] Maddox, R.N., Measuring satisfaction with tourism. *Journal of Travel Research*, **23**, pp. 2–5, 1985. DOI: 10.1177/004728758502300301.

[41] Williams, D. & Roggenbuck, J., Measuring place attachment: Some preliminary results, 1989.

[42]  Hair, J., Black, W. & Babin, B., *Multivariate Data Analysis*, 7th ed., Pearson Education, 2010.

[43]  Stylidis, D., The role of place image dimensions in residents' support for tourism development. *International Journal of Tourism Research*, **18**, pp. 129–139, 2016. DOI: 10.1002/jtr.2039.

# SECTION 4
# TOURISM AND
# SUSTAINABILITY

# ROLE OF ARTIFICIAL INTELLIGENCE AND BIG DATA ANALYTICS IN SMART TOURISM: A RESOURCE-BASED VIEW APPROACH

ASTERIOS STROUMPOULIS[1], EVANGELIA KOPANAKI[1] & SOTIRIOS VARELAS[2]
[1]Department of Business Administration, University of Piraeus, Greece
[2]Department of Tourism Studies, University of Piraeus, Greece

## ABSTRACT

This study examines the use and impact of artificial intelligence (AI) and big data analytics (BDA) in the tourism industry (TI). The digital age has brought a lot of changes transforming the business environment. The extensive use of the internet combined with the recent technological advances have greatly affected tourism companies, which need to face increased competition, changing tourists' needs and quick development of customer services. Furthermore, due to the widespread digitization, the tourism industry is overwhelmed by a huge amount of data that needs to be processed and analysed. AI is a rapidly evolving set of technologies, which can to some extent replace the analytical ability and decision-making capabilities of human beings. It can thus enable the development of innovative services and the intelligent processing of large amounts of data. Although AI is a widely known technology, it is still not widely used in the tourism industry. However, the adoption of AI is accelerated during the last 3 years, which is also reflected in the literature. By conducting an extensive literature review, this study aims to examine the level of adoption of AI applications in different sectors of the tourism industry and to discuss their role in big data analytics and in smart tourism (ST). It also aims to examine under which circumstances the adoption of these applications and technologies could enable tourism companies to obtain a competitive advantage. To explain this, the paper develops a conceptual framework using the resource based view theory. Based on the proposed framework, it shows that the adoption of the above combination may enable tourism companies to increase their business performance, achieve economic results and potentially attain a sustainable competitive advantage. Therefore, this research discusses the strategic role of AI and BDA in ST, makes propositions of implementation and forms the base for future research.

*Keywords: artificial intelligence, tourism industry, smart tourism, big data analytics, business performance, competitive advantage.*

## 1 INTRODUCTION

Many countries around the world have turned the tourism industry into a key economic factor, which contributes to their economic development and helps to increase the country's GDP. The use of information technology (IT) in the tourism industry has dynamically started two decades ago, where companies started to understand and realize the tremendous impact that IT could have at the industry, through the automation of many tasks and services [1]. New technologies and information systems provide companies with tools to find partners, offer products and manage better services for tourists [2].

The technological advances and the tremendous increase of IT solutions in the last decade, have been accelerated due to the expansion of the Internet and the development of social media [3]. These advances led to the digitization of tourists' services and companies, changing dramatically the environment of the tourism industry [4]. Moreover, the transition of websites into mobile applications, which enhance the participation of users, known as Web 2.0 applications, have led the entire tourism industry to the "informatization" of the value chain [5] and the increase of the concept of Travel 2.0 [6], which allows tourists to publish their experiences through internet sites about any travel-related content [7].

WIT Transactions on Ecology and the Environment, Vol 256, © 2022 WIT Press
www.witpress.com, ISSN 1743-3541 (on-line)
doi:10.2495/ST220091

Therefore, a huge amount of data is generated on a daily basis inside the tourism industry. The importance of the data and information handling for the entire industry has been documented by both practitioners and academics [8], [9]. The existence and use of information in the tourism sector brought not only many opportunities, but also threats for companies [10].

Artificial intelligence (AI) is identified as a disruptive factor in the current digital era. This paper tries to identify the theoretical and practical benefits that AI could bring in the tourism companies, especially in difficult periods, where companies of the tourism industry face situations leading to an economic stagnation. According to Xiang [11], AI is related to innovations and plays an important role in the tourism industry. It is thus expected to gradually change the services and experience of tourists. According to Tuo et al. [12], there are a few studies which examine the effects of AI in the tourism industry. The majority of studies do not analyse the characteristics of this technology and do not thoroughly analyse its impact. Furthermore, only limited studies analyse if the use of AI could enable companies to improve their position inside their task environment and increase their performance.

All these data which are produced, through the internet or various applications, must be analysed in order to help managers into the decision-making process. Therefore, this significant contribution of big data has led academics to explore the potential of the big data analytics (BDA) in the tourism industry [13]. BDA could help companies to better understand customers´ motivation through conducting online reviews [14] and analysing social media data. Moreover, BDA could contribute to analyse customer value and examine its impact on hotel performance [15]. Hence, big data is an important source of low-cost data, which could lead to trace tourists' movements, preferences, favourite points of interests, behaviours, future trends [16].

It is an undeniable fact that only a few existing studies systematically explain how artificial intelligence and big data analytics could affect the tourism industry and how these technologies could improve the internal operations of the companies in near the future. To address these issues, both from a practical and theoretical perspective, this paper proposes a conceptual framework on how AI and BDA could contribute to the tourism industry.

The current literature on these technologies and on their contribution to the tourism industry have not been thoroughly explored, despite the benefits that could derive by their use, to both managers and customers [17]. The purpose of this paper is to examine the level of adoption of AI applications in different domains of the tourism industry and to discuss their role in Big Data analytics and especially in sentiment analysis. An additional objective is to examine under which circumstances the adoption of AI applications could enable tourism companies to increase their performance or could additionally enable them to obtain competitive advantage.

Therefore, this is a conceptual paper, which aims to analyse the subject of study theoretically. Section 2 examines the concepts of smart tourism, AI and BDA, as analysed in the literature and describes how these technologies could contribute to the tourism industry. Moreover, this section describes the theory which is used as a base for the development of the conceptual framework. Section 3 presents and describes the proposed the framework. Finally, the conclusion and discussion section compares the results of the paper with those of the literature and summarizes the authors' contributions. It reveals the limitations of this paper and gives directions for future research.

## 2 LITERATURE REVIEW

Many sectors, such as the automotive and financial sectors, are early adopters of new technologies, such as artificial intelligence applications. In contrast, other sectors, such as education and tourism, are more hesitant to adopt new technologies. Hence, they remain less digitized than others [17]. Information systems and new technologies are very important for companies, as with their implementation, companies can create value and possibly gain competitive advantage. Therefore, information technology must be part of any company's business strategy [18].

Nowadays, the tourism industry produces a huge amount of data due to tourists' opinions, social media, flights information and hotel reservations etc. [19]. Despite this enormous production of data and the existence of new technologies, artificial intelligence has not proved yet its full potential [20], as its use is still not widespread.

This paper aims to clarify and analyse the impact of artificial intelligence technology and big data analytics in the tourism industry. It also aims to examine how the combination of these technologies could boost the tourist companies to increase their performance and/or obtain a competitive advantage.

To better understand and analyse the concepts addressed in this paper, it was necessary to conduct a literature review. The literature review is significant because it can provide information and results of previous research papers, as well as reveal the theories developed or used in the subject under examination. It also helps the researchers to support their argument, while also providing an original contribution [21].

The steps, that were followed in this research, include: searching for articles by keywords, narrowing the article selection by reading the abstracts, clarifying the meanings and the relationship among them, identifying the gap in the literature, discuss the results of this paper in relation to the literature.

The keywords that were used in the search were: artificial intelligence (AI), smart tourism, big data analytics, business performance. The search was conducted mostly in Google Scholar, Scopus and Researchgate databases.

The literature review enabled us to explain the main technologies, concepts and theories related to the subject of study. It also allowed us to develop a conceptual framework that presents the main factors enabling a tourism company to increase its performance.

## 2.1 Smart tourism

The concept of "smart tourism" was primarily introduced by IBM in 2008. Since then, the tourism industry has gradually started to embody technologies, such as internet applications, Internet of Things, cloud computing and AI technology [22]. According to Gretzel et al. [23, p. 181], smart tourism is defined as "tourism supported by integrated efforts at a destination to collect and aggregate/harness data derived from physical infrastructure, social connections, government/organizational sources and human bodies/minds in combination with the use of advanced technologies to transform that data into on-site experiences and business value-propositions with a clear focus on efficiency, sustainability and experience enrichment".

Smart tourism is a mix of three different elements, which are [23]:

- Smart destinations;
- Smart experience;
- Smart business.

Smart tourism is not only the collection of huge amounts of data, but also the storage, combination, analysis, and usage of them, in order to inform the operators, service providers and customers [18]. The goal of smart tourism is firstly to develop new information substructures by gathering all the information and secondly to improve the procedures of management, to promote innovation and increase the competitiveness of tourism companies and destinations [24].

As far as the tourism industry is concerned, there are six core resources: physiography, culture and history, tourism superstructure, market ties, mix of activities, and special events [25]. Due to the technological advances of the last decades these resources have changed to smart tourism resources [26]. Smart tourism has the ability to connect the digital world with the physical world during and after travelling [23]. Smart tourism could make the connection by predicting the needs of the customers, by improving the traveller's experience and by convincing travellers to share their experiences with others [23].

Smart tourism is constituted by specific core technologies, such as cloud computing, Internet of things, mobile terminal communication and artificial intelligence [22]. Artificial intelligence is the main technology of smart tourism and the main core of intelligent tourism. Therefore, AI could help tourism companies to use the huge amount of available information resources in order to increase their performance.

## 2.2  Artificial intelligence

Artificial intelligence is not a new concept and technology, because it was first proposed by John McCullach in 1955 [27], as a science which is able to make the machines smart. Artificial intelligence could be considered as the human intelligence, which is technology materialized by computer programs. So, artificial intelligence is developed and implemented by machines, in contrast to the natural intelligence which derives by humans and animals [22]. According to He [28] artificial intelligence could be classified in three categories: (1) weak AI; (2) strong AI; and (3) artificial super intelligence.

Artificial intelligence can be used to provide customers at any industry with personalized products or with interactive and unique experience. It can also effectively replace the manpower in customers services, manufacturing plants etc.

Despite the fact that artificial intelligence is an old concept, it is a new technological science area, with limited applications in the tourism industry. Digital transformation in the tourism sector is mostly concerned with the development of applications, such as mobile boarding systems and online check-in systems, which facilitate transactions and can be used by many customers and consumers [8]. However, during the last years, artificial intelligence is also gaining the trust of managers and is starting to become part of a company's routine [29]. According to Geisler [30], there are two main types of artificial intelligence, the pure digital ones, and the robots. Both types of artificial intelligence are gradually used in the hospitality and tourism industry [31].

## 2.3  Artificial intelligence in the tourism industry

The development and adoption of information technology has brought changes in the tourism and hospitality industry [31]. Artificial intelligence, as mentioned above, is a new form of intelligence which is able to synthesize different ideas at the same time [32]. Also, it can respond to any questions posed by customers and at the same time provide valuable information to tour operators and other companies. Moreover, as tourists and customers are becoming more and more demanding, expecting from companies to respond quickly to their needs, artificial intelligence could help companies to deliver immediate responses, without delays caused by the involvement of staff [31].

In this new era, where digitalization plays an important role in every person's life and company's function, artificial intelligence has gained customer's trust by offering convenient interaction both online and offline. The application of this technology gives customers the potential to obtain a more interactive experience (in terms of marketing) and to have a better shopping or booking experience, which could increase the customer's satisfaction, loyalty and consumption [33]. With these results artificial technology can support companies to increase their efficiency and productivity, as well as their economic performance [34]. So, the application of artificial intelligence will be soon expanded in many sectors and operations of the tourism industry.

There are several benefits, which can derive from the use of AI. As far as the financial results are concerned, AI can decrease the labour costs. Robots, chatbots and self-service kiosks are able to operate during a day, every day, without any physical support. They are also able to serve more customers than human employees, leading to better sales numbers [35].

Apart from the financial results, the use of this technology can also bring non-financial benefits to tourism companies. First of all, AI could provide consumers and travellers with more attractive and interactive applications to increase the level of their engagement [36]. Moreover, as mentioned before, robots, chatbots and kiosks, are able to communicate in any language, unlike humans, who are not able to do that [37]. In addition, these applications could save quality time for employees, who would be able to do more creative and important tasks inside the company. Last but not least, a company adopting this technology, would increase its reputation as one of the most high-tech companies inside the industry [35].

According to Samara et al. [17], apart from the above financial and non-financial benefits, artificial intelligence is also able to offer competitive advantage to a company in the tourism industry. As there is evidence that the adoption of this new technology from other companies inside the industry is limited, this means that any company, which is an early adopter, will be able to create a competitive advantage.

### 2.4 Big data analytics and smart tourism

According to Chen et al. [38], "big data" is a concept which describes datasets, which are so large, unstructured, and complex that they require advanced and unique technologies, in order to be stored, managed, analysed, and visualized. Nowadays, due to the internet, there is a huge amount of tourism information, such as, pre-trip planning, reservation, booking, reviews, comments, photos, experiences, social media interaction, etc. All this information becomes tourism big data, which need to be captured and analysed so as to uncover tourism trends, correlations and other insights inside the tourism sector [39].

The technological advances and the development of innovative services made tourists to seek for personalized services and products. The data deriving from this search and usage has proved to be one of the most significant steps for the development of smart tourism [40]. Therefore, smart tourism is closely related to the collection of large amounts of data, intelligently stored, processed, and analysed. Consequently, smart tourism relies on big data to ameliorate services and support the decision-making process of managers [41].

Furthermore, as the hospitality and tourism industry are vulnerable to many factors and environmental uncertainties, the use of data collection techniques, methods of statistical analysis and forecasting procedures are truly necessary [34] for their survival. Additionally, big data analysis seems to be the key for the development of smart tourism and the attainment of competitive advantage.

## 3 THEORETICAL DEVELOPMENT

Resource-based view (RBV) theory is a theory which analyses the internal resources and capabilities of a firm, supporting the creation of competitive advantage [42]. Therefore, RBV theory examines the resources of a company and analyzes their characteristics. It supports that the attainment of a sustainable competitive advantage could be created by the unique combination of resources at the core of the company [43]. Companies could achieve sustainable competitive advantage if they take advantage of their internal resources, such as human capital and information technology [44].

According to the RBV theory, competitive advantage can be additionally created by the development of resources, which have VRIO characteristics [45]. These characteristics include value, rareness, imitability and organization [46]. It can be noted that the VRIO characteristics can be developed beyond the boundaries of an organization, by combining the resources, which are available to different members of the tourism industry [47].

In their research, Stroumpoulis et al. [44], use the RBV theory to examine the impact of information technology on supply chain management. They propose a conceptual framework which shows how the combination of human, IT and other resources of a company could lead to specific capabilities and then to competitive advantage.

So, according to the RBV-theory and the above analysis of the literature, artificial intelligence applications and big data analysis can form unique resources that could enable a company to gain competitive advantage.

### 3.1 Conceptual framework

As mentioned above, information technology, including AI applications, forms an internal source of the company, which provides companies with important data and information. Moreover, in the tourism industry, if a company implements AI technology, it is considered as an early adopter, being at a beneficial position inside its task environment [17].

As analyzed in Section 2, the applications of AI could allow companies to collect the necessary information of what tourists like and dislike and to reveal the trends of travellers and consumers, so as to proceed to specific marketing strategies to specific target groups [48]. In addition, the analysis of these data, through BDA, could help managers to take the right decisions. Moreover, it could enable companies to design more personalised and customised experiences for each tourist, creating in this way an important value for them.

Therefore, according to the RBV-theory and the proposed framework by Stroumpoulis et al. [44], the implementation of AI technology and big data analytics in the tourism industry could enable companies to develop specific "smart capabilities". According to Debnath et al. [49], the above capabilities should provide companies at the basic levels with the functions of sensing, processing, controlling and communicating and at the advanced levels with the functions of predicting, healing and preventing. So, these capabilities could not be bought or transferred, but could only be gained through time and continuous use.

These capabilities would help companies to develop strategies in order to increase their business performance by improving the cost and time needed for their internal operations and customers' services. Therefore, they could increase their customers' loyalty and improve their economic results. Finally, these results could enable them to strengthen their position inside the industry and potentially obtain a sustainable competitive advantage.

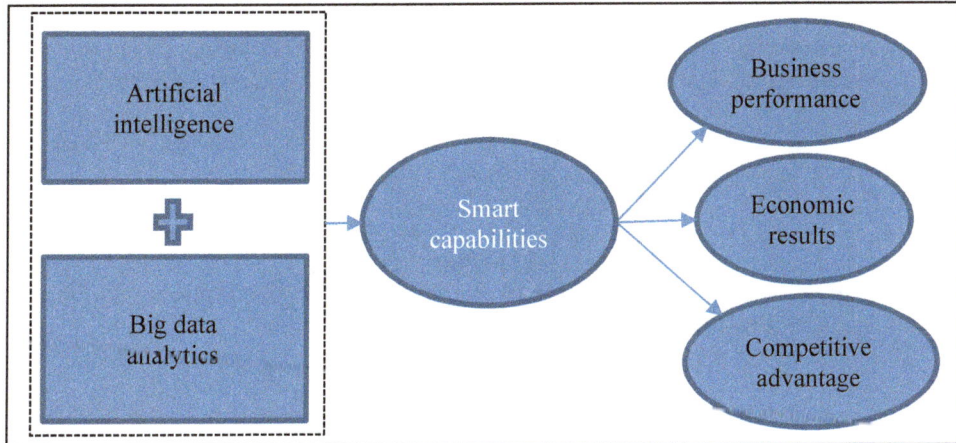

Figure 1:  Combining AI and BDA in tourism industry.

## 4  CONCLUSIONS

The aim of this research was to contribute to the literature, examining the impact of artificial intelligence and big data analytics on the tourism industry. An additional aim was to examine whether their combination could lead to improved business results. Artificial intelligence may provide many benefits both at the business and the customer (tourist) level. It can form the base for the development of applications, leading to services optimization and improved user experience. It can also contribute to the analysis of the large amount of data produced in the industry. Combined with big data analytics it can provide smart capabilities to companies leading them to improved decision making.

To better explain the impact of these technologies on the tourism industry, the paper developed a framework based on the RBV theoretical approach. The analysis showed that the use of these technologies enables tourism companies to increase their business performance, establish better relationships with their customers, achieve better economic results and potentially gain competitive advantage.

So, this paper provides the literature of smart tourism with a conceptual framework to evaluate the impact of AI applications and big data analysis. However, due to the fact that AI is a new technology, the level of adoption in the industry is still very low, apart from specific endeavours, mostly of big companies. Therefore, there is still much room for IT research in smart tourism, which could be used by companies to ensure their benefits in their task environment. Hence, a significant limitation of this area of research, is that academics are still not able to use large scale surveys in the industry, in order to verify their theories and frameworks.

In conclusion, although the potential impact of artificial intelligence and big data analytics is discussed in the literature and presented in the proposed conceptual framework, companies inside the industry have not yet widely adopted these technologies. However, understanding the impact and the full potential of these technologies is of major importance for companies, as their development and adoption will probably be the next big thing in the tourism industry.

## ACKNOWLEDGEMENTS

This work has been partly supported by the University of Piraeus Research Center.

## REFERENCES

[1] Law, R., Buhalis, D. & Cobanoglu, C., Progress on information and communication technologies in hospitality and tourism. *International Journal of Contemporary Hospitality Management*, **26**(5), pp. 727–750, 2014.

[2] Buhalis, D. & O'Connor, P., Information communication technology revolutionizing tourism. *Tourism Recreation Research*, **30**(3), pp. 7–16, 2005.

[3] Chuang, T.C., Liu, J.S., Lu, L.Y., Tseng, F.M., Lee, Y. & Chang, C.T., The main paths of eTourism: Trends of managing tourism through internet. *Asia Pacific Journal of Tourism Research*, **22**(2), pp. 213–231, 2017.

[4] Ashari, H.A., Heidari, M. & Parvaresh, S., Improving SMTEs' business performance through strategic use of information communication technology: ICT and tourism challenges and opportunities. *International Journal of Academic Research in Accounting, Finance and Management Sciences*, **4**(3), pp. 1–20, 2014.

[5] Cluster, W.D., Pardo, P.D., Cooper, M. & Tajeddini, K., Tourism and social media. *Encyclopedia of Information Science and Technology*, 3rd ed., IGI Global, pp. 3652–3665, 2015.

[6] Oklobdžija, S.D. & Popesku, J.R., The link between digital media and making travel choices. *Marketing*, **48**(2), pp. 75–85, 2017.

[7] Muñoz-Leiva, F., Hernández-Méndez, J. & Sánchez-Fernández, J., Generalising user behaviour in online travel sites through the Travel 2.0 website acceptance model. *Online Information Review*, **36**(6), pp. 879–902, 2012.

[8] Gretzel, U., Intelligent systems in tourism: A social science perspective. *Annals of Tourism Research*, **38**(3), pp. 757–779, 2011.

[9] Koo, C., Gretzel, U., Hunter, W.C. & Chung, N., Editorial: The role of IT in tourism. *Asia Pacific Journal of Information Systems*, **25**(1), pp. 99–102, 2015.

[10] Xiang, Z., Magnini, V.P. & Fesenmaier, D.R., Information technology and consumer behavior in travel and tourism: Insights from travel planning using the internet. *Journal of Retailing and Consumer Services*, **22**, pp. 244–249, 2015.

[11] Xiang, Z., Tourism in the era of artificial intelligence. *J Tour*, **35**(01), pp. 1–3, 2020.

[12] Tuo, Y., Ning, L. & Zhu, A., How artificial intelligence will change the future of tourism industry: The practice in China. *Proceeding of the ENTER 2021 eTourism Conference*, pp. 83–94, 2021.

[13] D'Acunto, D. & Volo, S., Cultural traits in the consumption of luxury hotel services. *Proceeding of the ENTER 2021 eTourism Conference*, pp. 269–279, 2021.

[14] Yoo, K.H. & Gretzel, U., What motivates consumers to write online travel reviews? *Information Technology and Tourism*, **10**(4), pp. 283–295, 2008.

[15] Xie, K.L., Zhang, Z. & Zhang, Z., The business value of online consumer reviews and management response to hotel performance. *International Journal of Hospitality Management*, **43**, pp. 1–12, 2014.

[16] Yang, W. & Mattila, A.S., Why do we buy luxury experiences? Measuring value perceptions of luxury hospitality services. *International Journal of Contemporary Hospitality Management*, **28**(9), pp. 1848–1867, 2016.

[17] Samara, D., Magnisalis, I. & Peristeras, V., Artificial intelligence and big data in tourism: A systematic literature review. *Journal of Hospitality and Tourism Technology*, **11**(2), pp. 343–367, 2020.

[18] Tsaih, R.H. & Hsu, C.C., Artificial intelligence in smart tourism: A conceptual framework. *Proceedings of the 18th International Conference on Electronic Business*, pp. 124–133, 2018.

[19]    Guan, D. & Du, J., Cross-media big data tourism perception research based on multi-agent. *Proceedings of the 2015 Chinese Intelligent Systems Conference*, pp. 353–360, 2016.

[20]    Talwar, R. & Koury, A., Artificial intelligence: The next frontier in IT security? *Network Security*, **2017**(4), pp. 14–17, 2017.

[21]    Pozzebon, M., Petrini, M., de Mello, R.B. & Garreau, L., Unpacking researchers' creativity and imagination in grounded theorizing: An exemplar from IS research. *Information and Organization*, **21**(4), pp. 177–193, 2011.

[22]    Zhang, L. & Sun, Z., The application of artificial intelligence technology in the tourism industry of Jinan. *Journal of Physics: Conference Series*, **1302**(3), 032005, 2019.

[23]    Gretzel, U., Sigala, M., Xiang, Z. & Koo, C., Smart tourism: Foundations and developments. *Electronic Markets*, **25**(3), pp. 179–188, 2015.

[24]    Shafiee, S., Ghatari, A.R., Hasanzadeh, A. & Jahanyan, S., Developing a model for sustainable smart tourism destinations: A systematic review. *Tourism Management Perspectives*, **31**, pp. 287–300, 2019.

[25]    Crouch, G.I. & Ritchie, J.B., Tourism, competitiveness, and societal prosperity. *Journal of Business Research*, **44**(3), pp. 137–152, 1999.

[26]    Sigala, M., New technologies in tourism: From multi-disciplinary to anti-disciplinary advances and trajectories. *Tourism Management Perspectives*, **25**, pp. 151–155, 2018.

[27]    McCullagh, J., Bluff, K. & Ebert, E., A neural network model for rainfall estimation. *Proceedings of the 1995 Second New Zealand International Two-Stream Conference on Artificial Neural Networks and Expert Systems*, pp. 389–392, 1995.

[28]    He, Z., Social transformation and administrative ethics in the era of artificial intelligence: Can machines manage people? *Electron. Gov.*, **11**, pp. 2–10, 2017.

[29]    Riccio, F., Vanzo, A., Mirabella, V., Catarci, T. & Nardi, D., Enabling symbiotic autonomy in short-term interactions: A user study. *Proceedings of the International Conference on Social Robotics*, pp. 796–807, 2016.

[30]    Geisler, R., Artificial intelligence in the travel and tourism industry adoption and impact. Doctoral dissertation, Repositorio UniversidadeNova, Portugal, 2018.

[31]    Popesku, J., Current applications of artificial intelligence in tourism and hospitality. *Proceedings of the International Scientific Conference on Information Technology and Data Related Research*, pp. 84–90, 2019.

[32]    Zsarnoczky, M., How does artificial intelligence affect the tourism industry? *VADYBA*, **31**(2), pp. 85–90, 2017.

[33]    Luo, X., Tong, S., Fang, Z. & Qu, Z., Frontiers: Machines vs. humans: The impact of artificial intelligence chatbot disclosure on customer purchases. *Marketing Science*, **38**(6), pp. 937–947, 2019.

[34]    Kazak, A.N., Chetyrbok, P.V. & Oleinikov, N.N., Artificial intelligence in the tourism sphere. *IOP Conference Series: Earth and Environmental Science*, **421**(4), 042020, 2020.

[35]    Ivanov, S.H. & Webster, C., Adoption of robots, artificial intelligence and service automation by travel, tourism and hospitality companies: A cost–benefit analysis. Artificial intelligence and service automation by travel, tourism and hospitality companies: A cost–benefit analysis. *Proceedings of the International Scientific Conference: Contemporary Tourism – Traditions and Innovations*, 2017.

[36]    Kuo, C.-M., Chen, L.-C. & Tseng, C.-Y., Investigating an innovative service with hospitality robots. *International Journal of Contemporary Hospitality Management*, **29**(5), pp. 1305–1321, 2017.

[37] Saber Chtourou, M. & Souiden, N., Rethinking the TAM model: Time to consider fun. *Journal of Consumer Marketing*, **27**(4), pp. 336–344, 2010.

[38] Chen, H.C., Chiang, R.H.L. & Storey, V.C., Business intelligence and analytics: From big data to big impact. *MIS Quarterly,* **36**(4), pp. 1165–1188, 2012.

[39] Liu, Y.Y., Tseng, F.M. & Tseng, Y.H., Big data analytics for forecasting tourism destination arrivals with the applied vector autoregression model. *Technological Forecasting and Social Change*, **130**, pp. 123–134, 2018.

[40] Xiang, Z. & Fesenmaier, D.R. (eds), Big data analytics, tourism design and smart tourism. *Analytics in Smart Tourism Design: Concepts and Methods*, Springer: Germany, pp. 299–307, 2017.

[41] Park, S., Big data in smart tourism: A perspective article. *Journal of Smart Tourism,* **1**(3), pp. 3–5, 2021.

[42] Barney, J., Firm resources and sustained competitive advantage. *Journal of Management,* **17**(1), pp. 99–120, 1991.

[43] Conner, K. & Prahalad, C., A resource-based theory of the firm: Knowledge versus opportunism. *Organisation Science*, **7**(5), pp. 477–501, 1996.

[44] Stroumpoulis, A., Kopanaki, E. & Karaganis, G., Examining the relationship between information systems, sustainable SCM, and competitive advantage. *Sustainability*, **13**(21), 11715, 2021.

[45] Hart, S.L., A natural-resource-based view of the firm. *The Academy of Management Review*, **20**(4), pp. 986–1014, 1995.

[46] Cardeal, N. & Antonio, N.S., Valuable, rare, inimitable resources and organization (VRIO) resources or valuable, rare, inimitable resources (VRI) capabilities: What leads to competitive advantage? *African Journal of Business Management*, **6**(37), pp. 10159–10170, 2012.

[47] Dyer, J.H. & Singh, H., The relational view: Cooperative strategy and sources of interorganizational competitive advantage. *Academy of Management Review*, **23**(4), pp. 660–679, 1998.

[48] Fatima, S., Desouza, K.C. & Dawson, G.S., National strategic artificial intelligence plans: A multi-dimensional analysis. *Economic Analysis and Policy*, **67**, pp. 178–194, 2020.

[49] Debnath, A.K., Chin, H.C., Haque, M.M. & Yuen, B., A methodological framework for benchmarking smart transport cities. *Cities, The International Journal of Urban Policy and Planning*, **37**, pp. 47–56, 2014.

# IN SEARCH OF THE DESIRED SUSTAINABLE TOURISM: A REVIEW OF LIFE CYCLE ASSESSMENT (LCA) TOURISM STUDIES

CRISTINA CAMPOS HERRERO[1], JARA LASO[1], PERE FULLANA-I-PALMER[2], JAUME ALBERTÍ[2],
MARGALIDA FULLANA[2], ÁNGEL HERRERO[3], MARÍA MARGALLO[1] & RUBÉN ALDACO[1]
[1]Department of Chemical and Biomolecular Engineering, University of Cantabria, Spain
[2]UNESCO Chair in Life Cycle and Climate Change ESCI-UPF, Spain
[3]Department of Business and Administration, University of Cantabria, Spain

## ABSTRACT
Sustainable tourism should be promoted as a new system for the sustainable management of resources from a socioeconomic and environmental point of view. For this purpose, it is necessary to develop a tool capable of assessing the impacts associated with each of the stages of the sector and to identify which actions are currently being addressed in the tourism sector in order to achieve the desired sustainability in the sector. This timely study aims to describe the current framework of life cycle assessment (LCA) and its application to the tourism sector. To address these questions, the geographical distribution, the temporal evolution of the publications, as well as the most relevant characteristics of the tourism industry articles were evaluated such as the functional unit and system boundaries considered. The study identifies key recommendations on the progression of LCA for this increasingly important sustaining tourism sector. As important results, it stands out that 94% of articles focused on LCA methodology were from the last decade and almost 26% of the articles reviewed cover sustainable tourism term, considering environmental, social and economic aspects. Specifically, LCA is a highly effective tool capable of assessing direct and indirect carbon emissions at all stages of the activity as well as the socioeconomic and environmental impacts generated in the tourism sector. This review showed that the most common environmental indicator in the LCA methodology is the carbon footprint. COVID-19 pandemic is also an object of discussion in the framework of the sustainable tourism together with advocating support for the eco-labelling and digitalisation of the tourism experiences as valuable tools to minimize environmental negativities, to promote mechanisms to access green markets and to frame successful synergies.
*Keywords: life cycle assessment, sustainable tourism, eco-labelling, COVID-19, carbon footprint.*

## 1 INTRODUCTION
Tourism is one of the most important sectors in the global economy, which has been growing steadily over the recent decades and whose forecasts for the coming years predict a strengthening of this trend [1]. This competitive and dynamic industry on a global scale, employs millions of people, moves billions of dollars, and generates and induces technological innovation. In fact, in 2019, the tourism industry generated more than USD 236 billion, surpassing oil exports and food production in terms of business turnover) [2]. However, 2020 and 2021 have been the worst years on record for tourism, which has been one of the hardest hit economic sectors by the COVID-19 pandemic, facing a decline in international tourist arrivals during 2020 of between 58% and 78%, with a drop in direct tourism employment of between 100 and 120 million people. In addition, in 2020 international tourism profits fell by 64% in real terms (local currencies, constant prices), equivalent to a decrease of more than USD 900 billion, cutting the overall value of world exports by more than 4% in 2020. The total loss of international tourism export earnings amounts to almost US$ 1.1 trillion. Asia–Pacific (–70% in real terms) and the Middle East (–69%) recorded the largest revenue declines [3]. Then, in the first quarter of 2021, international arrivals were 83% lower (180 million fewer arrivals).

WIT Transactions on Ecology and the Environment, Vol 256, © 2022 WIT Press
www.witpress.com, ISSN 1743-3541 (on-line)
doi:10.2495/ST220101

Indeed, tourism is today's most significant and important industry [4], with the greatest environmental impact. Between 2009 and 2013, the global carbon footprint (CF) of tourism increased from 3.9 to 4.5 gigatonnes of carbon dioxide equivalent and this growth accounted for around 8% of global carbon emissions, with transport, shopping and food being the main contributors [5]. Around 2.4% of global $CO_2$ emissions come from aviation, being one of the most polluting activities in tourism [6]. Despite all this, one positive outcome of the pandemic crisis was a reduction in emissions and improvements in air quality. Furthermore, global carbon emissions in 2020 are estimated to have fallen by 8% in tourism sector [7]. Therefore, the need to transform the operation of the tourism sector towards sustainable tourism remains indispensable for the sector to continue to grow towards international targets and a great opportunity for the tourism sector [8].

Sustainable tourism is widely recognized as the only solution to large-scale tourism's negative effects [9]. According to United Nations Environment Programme (UNEP) and the United Nations World Tourism Organization (UNWTO), sustainable tourism is "the development of tourism activities with an appropriate balance between environmental, economic and sociocultural dimensions to ensure their long-term sustainability". Furthermore, it should satisfy the needs of existing tourists and destinations while providing opportunities for further development in the future, as well as maintaining heritage integrity, ecological integrity, biological diversity and livelihood system. In fact, tourism development should refer to sustainable development that aims at continuous improvement of tourist satisfaction [9]. A holistic balance between three dimensions (environmental, economic, socio-cultural) must be considered to try to achieve globally accepted sustainable tourism so as to ensure the short and long-term sustainability of the tourism sector in the face of climate change [10]. These three dimensions have key elements such as ecotourism, rural tourism, cultural tourism (heritage), community tourism, as well as policies that implement the circular economy in sustainable tourism [11]. Tourism has been identified by UNEP as one of the ten economic sectors capable of contributing to the transition to a sustainable and inclusive green economy [12].

Emphasis has been given to existing research on the carbon footprint of tourism, as climate change has recently become a key issue on the international tourism policy agenda [13]. But the question is how to measure and reduce the environmental burdens of this sector. Recently, researchers, organisations and policy makers are striving to develop concepts and metrics that measure environmental sustainability. Among these concepts and metrics, life cycle assessment (LCA) is a methodology for quantifying the environmental impacts of a product, process, or service over the course of its entire life cycle [4]. This tool could provide a consistent analytical framework and environmental data support for decision-making, allowing for the development of sustainable solutions to global tourism challenges and the promotion of mechanisms that allow different tourism services to access green markets as well as its efficacy in identifying opportunities for improving environmental performance and defining sustainability strategies for tourism [14]. Currently, very few articles have focused exclusively on LCA methodology and tourism, as most of them are combined with other methods. Some studies such as Maugeri et al. [15] assessed by LCA a trip and overnight stay in a hotel during mid-season with the arrival and departure of the tourist at Fontanarossa Airport in Catania, Sicily. Others like De Camillis et al. [16] use the LCA methodology in order to assess the environmental impacts in a three-star hotel located in Pescara, Italy. In this paper, a quantification of current LCA studies in the tourism sector is given. From all the environmental indicators and impact categories, global warming potential has been widely analyzed, being used as a proxy for the entire set of impact categories [17]. Fig. 1 displays

WIT Transactions on Ecology and the Environment, Vol 256, © 2022 WIT Press
www.witpress.com, ISSN 1743-3541 (on-line)

the conceptual aspects required for achieving sustainable tourism as well as the carbon emissions associated with tourism before and after the COVID-19 pandemic.

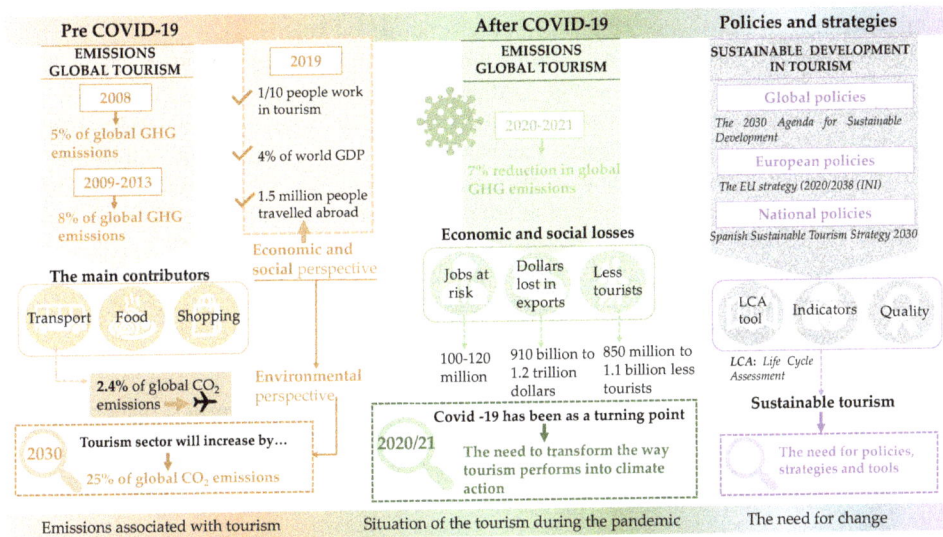

Figure 1:   Conceptual aspects for sustainable tourism. Greenhouse gas (GHG) emissions associated with tourism before and after the pandemic and policies and strategies for achieving sustainable tourism, being GHG and gross domestic product.

In this context, the main objective of this paper is to review the impact assessment methods and impact categories evaluated to the most appropriate combination to evaluate the environmental performance of tourism. As a result of the review of the articles, the most appropriate system is identified, which contributes to and advances the development of future environmental tourism saving the process of studying and selecting the optimum model. Moreover, as a secondary objective, this systematic review intends to contribute to a better understanding of sustainable tourism by compiling articles based not only on environmental but also on socio-economic aspects. To the best of our knowledge, the awareness created by COVID-19 and the climate crises have served as a new opportunity to change the current tourism model towards sustainable tourism.

## 2  MATERIALS AND METHODS

### 2.1  Literature search strategy and inclusion criteria

This review seeks to address the most relevant studies based on LCA methodology with the purpose of assessing environmental impacts and achieving global sustainable tourism. Searches of different sources of scientific literature, books and reports were included. Also, the Scopus database and other tourism sector specific databases such as the UNWTO were accessed. In addition, Google Scholar was also reviewed as a search engine. The definition of the scope made it possible to eliminate those documents that did not fit the object of the review. The review excluded studies that did not address tourism and that did not apply any type of environmental methodology such as LCA. Also, they were excluded those articles dealing with social issues, as tourism hospitality [18], sustainability of cities [19] or studies focused exclusively on tourist data [10]. In the same way, we discarded studies that used LCA methodology but they are focused on other sectors such as transport or buildings [20], or on other aspects like the use of plastic bottles in different cities [21].

For the bibliographic search, a classification of the documents was made by considering the LCA approach in the tourism sector. Furthermore, the CF indicator was considered as an environmental indicator due to its wide use in this field. The literature review was performed through a precise search in the Scopus database. Likewise, a bibliographic search was also made of those documents that deal tourism and LCA tool and CF indicator to the health pandemic suffered in 2020. Websites of international governmental organizations such as the UNWTO, the United Nations Conference on Trade and Development (UNCTAD), and the United Nations Economic Commission for Latin America and the Caribbean (ECLAC) and United Nations Organization (UN) were also used. Fig. 2 shows an overview of the steps to identify and classify the studies.

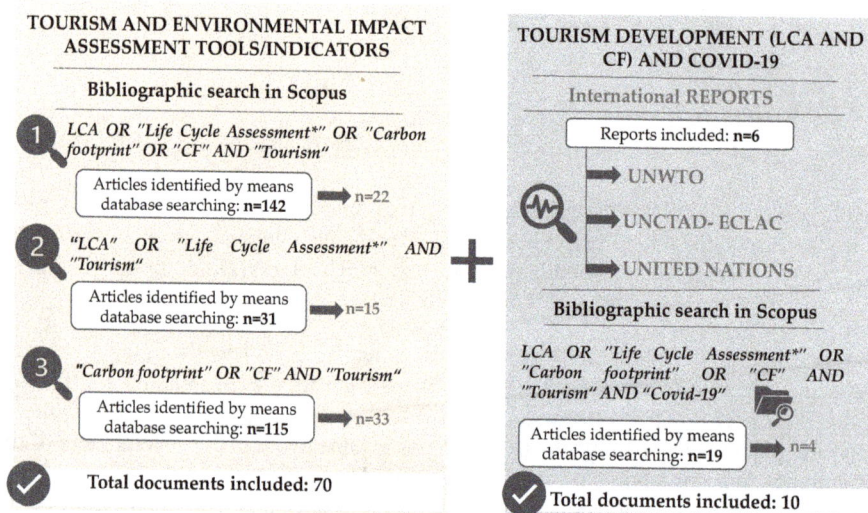

Figure 2: Schematic diagram of the steps followed in the bibliographic search.

## 2.2 Analysis of study findings

Each article of this review was assessed independently to identify the objective, the methodology followed and the conclusions. The following characteristics were analyzed in each article: phases of the system boundaries which leads to a delimitation of the different processes of the system under examination [22], functional unit (FU), the method used and impact categories studied. According to De Camillis et al. [23], these characteristics should be considered when developing this type of studies. The 4 characteristics analysed were:

(i) The choice of system boundaries so as to determine which process units will be included in the LCA study [24]. There are three options in this case: (a) *cradle to gate*, which considers stages from raw material extraction to transport use, accommodation, restaurants and leisure activities [25], (b) *cradle to grave*, which also examines the final disposal of waste [26], and (c) separate systems such as air transport used in tourism, hotels as an example of accommodation, restaurants or sport tourism as a leisure activity. The first two options consider a complete travel package while the third one examines only one of the sectors.

(ii) The second point to take into account is the FU, which is the reference unit on which all inputs and outputs of the system are based [24]. The FU for transport and tourism activities are well established and they can be defined as "1 passenger per kilometre driven" and "1 visitor activity performed" [27]. However, for a hotel stay, no consensus has yet been reached. Some studies use "per guest night" [28] following the PAS 2050:2011 standards. And in this case the carbon intensity of hotel operations is described on a "per capita" or "per user" basis. Another option is to consider the gross floor area of the hotel, such as "$m^2$ of the floor area" which considers energy and CF analysis [29]. On the other hand, if the whole trip is considered, the most common FU are "per trip" [30] or "one week of a holiday" [31]. In a large number of studies, the FU is not clearly defined and hinders the impact assessment process.

(iii) The third point is to consider the impact method applied. There are several methods for the LCA of tourism among which the CML-2001 developed by the University of Leiden stands out. This method groups life cycle inventory (LCI) results into midpoint categories by themes (e.g. climate change or ecological toxicity) [32]; the Eco-Indicator 99 developed by PRé Consultants that considers the environmental damage in human health, ecosystem quality and resources [33]; or the EDIP 1997 developed by the Institute for Product Development (IPU) at the Technical University of Denmark that uses a midpoint approach [32]. In other cases, the CF (an environmental indicator that measures total greenhouse gases) or the ecological footprint (EF) (based on resource consumption and waste production) [34] are used as tools to assess the impacts of tourism. There are also other environmental indicators such as Water Footprint (WF), Ecological Footprint (EF), the DEFRA (Food and Rural Affairs) [35], the TECI (Tourism Environmental Composite Indicator) [36] the TCQGMA (Environmental Management Module) [37] and the eco-efficiency model [38]. These indicators are more specific (in terms of the scope of geographical areas and tourism sub-sectors) and not as generic as the CF, so the use of these indicators is more limited in the tourism sector. In addition to these environmental methods, others consider economic and statistical aspects such as Input–Output analysis (IO) and Tourism Satellite Account (TSA) that can be combined with LCA or CF [39].

(iv) Finally, in relation to impact categories, the most frequently selected impact category in tourism LCAs is usually Global Warming Potential (GWP) in units of $CO_2$ eq. as this impact category is closely associated with the CF indicator, which is one of the important points of study in this review. The GWP impact category is sometimes combined with other impact categories such as acidification, eutrophication or ecotoxicity, depending on the scope of the study [40].

## 3 RESULTS AND DISCUSSION

### 3.1 Mapping and time evolution of the studies

A total of 80 documents were reviewed; 70 articles related to the tourism sector using LCA tool and/or CF environmental indicator, four articles addressing the situation of the tourism sector following these methodologies during the COVID-19 crisis and six reports.

Fig. 3 shows the location of the research institutions of the different studies that participated in the evaluated studies. Only the original articles were included in this figure, except for Lenzen et al. [5] because it refers to many countries and not one or two specific ones The reports were also not included in the figure taking into account they refer to large regions rather than specific countries. The icons represent the countries in which only LCA approach has been developed in the tourism sector, along with the number of studies conducted, while the coloured areas illustrate the territories where CF studies, or LCA and CF combined studies related to their application were conducted. Studies addressing the development of LCA tool focused on Asia (50%), Europe (43%) and Oceania (7%). Most of the articles were developed by institutions from a particular country, while four of them had international collaboration and involved researchers from four or five regions. This highlights the great importance of achieving global sustainable tourism and the need for a tool such as LCA to achieve this. It can be seen from the figure that the objective of using different tools to achieve sustainability in the sector is given for developed countries but not for developing countries, such as Africa. In the European context, Spain ranks first in the dissemination of studies using both LCA and CF approaches, producing or collaborating in nine publications. Italy (five studies), the United Kingdom (UK) (5) and Greece (4) also played an important role in the creation and study of tourism impact assessment tools. However, this is also paradoxical since the environmental impact of tourism is not analysed in the markets with the greatest potential impact, such as France, which is the world's leading tourist destination. Similarly, the United States, which is the third largest tourist destination in the world, does not present any of the studies on LCA in tourism [41].

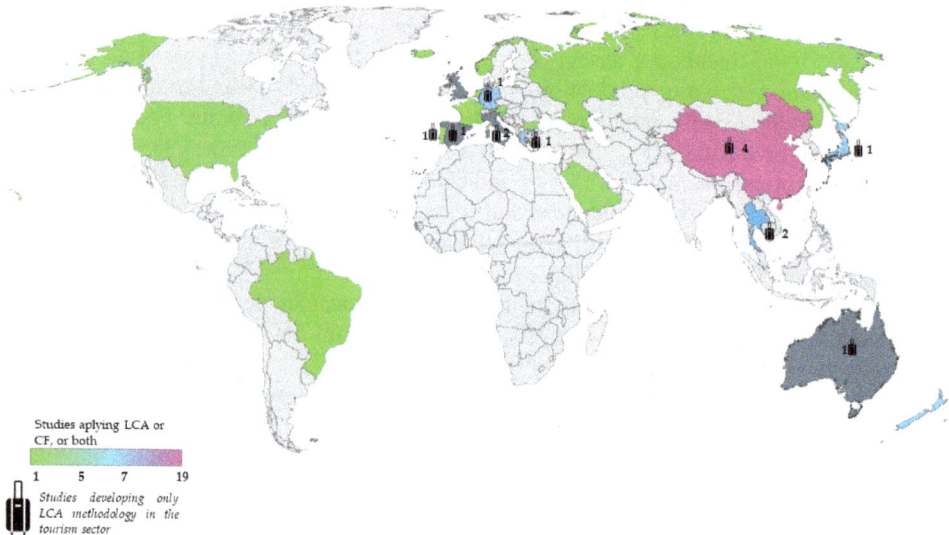

Figure 3:  Map showing the geographical distribution of studies included in the review.

According to the evolution of time of publications that use LCA and CF tools in the tourism sector, the period of time studied was from 2004 to 2021, i.e. for 17 years. The starting year chosen was 2004 since it is the first year in which there are peer-reviewed articles on tourism and LCA [23]. It can be seen that very few articles have been developed during the period (2004–2010). However, the development of this methodology has progressively increased, presenting a greater constancy and periodicity over the years, highlighting the year 2010 with six articles (three exclusively on LCA); 2016, with seven publications (two focused only on LCA) and 2020, with 15 (three focused only on LCA). The year 2021 (until June 2021) presents five articles (one exclusively on LCA) so far but an increase in publications is predicted with respect to 2020 given the growth of the sector and the concern to reduce its negative environmental impact. Ultimately, an increasing number of publications addressing LCA tool in tourism was observed in the period (2017–2021). Of the 74 articles in total about tourism, 37 were LCA or LCA-CF and 94% of them were from the last decade (2010–2021).

## 3.2 Implementation of sustainable tourism

Tourism is a human activity that involves the economy, the environment and society. Therefore, the objective will be to find the optimal compromise between environmental, economic and social variables in a defined time and space. Without sustainable tourism, there is a risk of entering a vicious circle in which biodiversity is lost, jobs and wealth are lost and there are demands in other markets [10]. In the review conducted in this paper, all the studies include the environmental variable but not all of them consider the social and economic variable. For this reason, it has been studied which articles take into account the term sustainable tourism when considering the three areas, only two of them or only the environmental variable (Fig. 4).

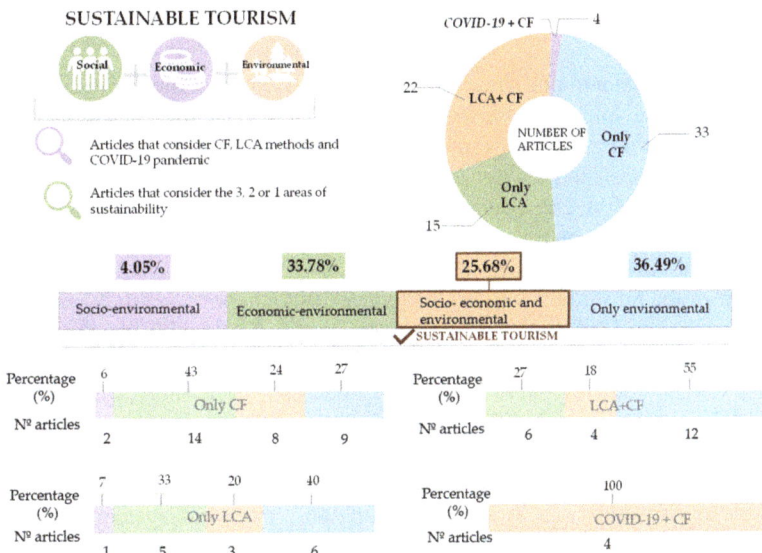

Figure 4:   Degree of implementation of sustainable tourism in the articles studied (74) in terms of the economic, social and environmental variables.

Firstly, of the 33 articles that study only the CF environmental indicator, 14 articles consider a socio-economic view of tourism, which represents 42% of the CF articles. For example, Sun [42] uses the TSA (Tourism Satellite Account) method that defines tourism expenditure in Taiwan (economic model) and the Environmentally Extended Input–Output Model (EEIO) to assess the CF (environmental indicator). Only 7% of the studies i.e. one article, they include a social and environmental perspective, such as a study in which tourists are evaluated through different questionnaires on some of the practices they perform [43]. According to the articles that considered both LCA tool and CF environmental indicator, it is noticed that 55% (12 articles) evaluate the studies from a unique environmental perspective [25], [26], [44], [45], 27% of the case studies, i.e. six articles, consider both economic and environmental perspective [46]–[48]. Finally, 18% considered the concept of sustainable tourism, as they study the three strands of sustainability as is the case of Gallucci and Dimitrova [49] who evaluate the CF of a city in Bulgaria, analyse social indicators and marketing strategies are suggested. Also, Cadarso et al. [50] which calculates the $CO_2$ emissions of an example of the Spanish tourism sector, uses a life cycle assessment input–output (LCA-IO) model and considers the social activity of the destination.

There are 15 articles considered the LCA tool, without combining it with the CF indicator, 40% considered only the environmental area as in the Cerutti et al [17] and Li et al. [39] studies, among others. 33%, i.e. five articles considered the environmental and economic aspects such as Gössling et al. [38] or König et al. [20] three articles in this group accounted for sustainability in tourism such as the study by Sun et al. [48]. Finally, only one article addressed social and environmental variables together [43].

Finally, the four articles studying the CF of the tourism sector during the health pandemic (CF and COVID-19) can be expected to consider the economics of the sector and how it has affected society and the environment. This is indeed the case, 100% of the articles have a sustainable tourism vision [52]–[55].

## 4 CONCLUSIONS

The tourism sector faces a major challenge in reducing GHG emissions due to the insatiable demand for travelling and the industry's desire to stimulate that demand. In response, researchers, organisations and policy makers are striving to develop concepts and metrics to measure environmental sustainability. Among those concepts and metrics, LCA is one of the most promising tools that can solve some of the drawbacks of existing environmental approaches and it has become a key target for tourism, as shown by the growing number of studies on tourism in recent years.

The review of the 80 studied showed that, since 2004, the increase in the development of articles employing the LCA tool has grown very successfully, with the progression being almost exponential from 2017 until now. Specifically, it stands out that 94% of articles focused on LCA methodology were from the last decade. Furthermore, the regions in which this methodology was considered were limited to developed countries, with a large proportion of studies focusing on Asia, with the great majority located in China, followed by territories in Europe and Oceania, giving an insight into the degree of concern and awareness of sustainable tourism and the LCA tool in these regions. In terms of systems characterisation, most of the articles addressed the accommodation, restaurant and transport stages in the study phases of the tourism sector. In addition, the FU most commonly used in the studies was "per guest night" or "per visitor" although this may change depending on the objective and scope of the study. By measuring environmental burdens and providing a reliable assessment of GHG emissions associated with tourist accommodation, a CF-LCA approach could be a scientific supporting tool for environmental communication and education of tourists, as well as an objective instrument for more responsible consumption. Likewise, GWP was the most studied impact category due to its strong influence and importance in the sector. Furthermore, in the last year, there has been an increase in the number of studies that take into account the impact of COVID-19 and it is considered a turning point for the reduction of carbon emissions for the integration of policies and strategies in the framework of sustainable tourism.

Answering to the broad paradigm of sustainability, this paper quantifies the articles that consider the three dimensions of sustainability. Specifically, 26% of the articles studied already apply the commitment between environmental, economic and social variables that is optimal in a defined space and time frame in order to achieve sustainable tourism as marked by international institutions such as UNWTO and other governmental entities.

Finally, this review also examines some of the strategies that are currently being implemented to achieve sustainability in tourism, such as the use of eco-labelling, in order to know its acceptable level of environmental impact, digitalisation and good practices by tourists. These strategies will help both public administration and tourists to make more sustainable tourism choices. Definitely, the awareness created by the COVID-19 crisis and the climate crisis can be considered an opportunity to take these measures.

Further work is recommended and it can be oriented to study in depth the socio-economic variable of the tourism sector in order to have a more detailed knowledge of the models used in these areas of sustainability.

## ACKNOWLEDGEMENTS

This research was funded by the INTERREG SUDOE Programme, grant number GREENTOUR: Circular Economy and Sustainable Tourism in Destinations of the SUDOE space (SOE4/P5/E1089). Furthermore, the authors are grateful to the anonymous reviewers whose comments and corrections have significantly improved the quality of this contribution.

## REFERENCES

[1]   Raggi, A. & Petti, L., A newly developed integrated environment quality approach for the design of hotel services. Progress in Industrial Ecology. *An International Journal*, **3**(3), pp. 251–271, 2006.

[2]   UNWTO, Tourism highlights: 2018 edition, 2018. https://www.e-unwto.org/doi/book/10.18111/9789284419876. Accessed on: 20 Apr. 2022.

[3]   UNWTO, World tourism barometer and statistical annex: Vaccines and reopen borders driving tourism's recovery, 2021. https://www.unwto.org/taxonomy/term/347. Accessed on: 19 Apr. 2022.

[4]   Puig, R., Kiliç, E., Navarro, A., Albertí, J., Chacón, L. & Fullana-i-Palmer, P., Inventory analysis and carbon footprint of coastland-hotel services: A Spanish case study. *Sci. Total Environ.*, **595**, pp. 244–254, 2017.

[5]   Lenzen, M., Sun, Y.Y., Faturay, F., Ting, Y.P., Geschke, A. & Malik, A., The carbon footprint of global tourism. *Nat Clim Chang.*, **8**(6), pp. 522–528, 2018.

[6]   Timperley, J., Should we give up flying for the sake of the climate? 2020. https://www.bbc.com/future/article/20200218-climate-change-how-to-cut-your-carbon-emissions-when-flying. Accessed on: 20 Apr. 2022.

[7]   International Energy Agency, Global energy review 2020: The impacts of the COVID-19 crisis on global energy demand and $CO_2$ emissions. Report, 2020.

[8]   World Tourism Organization and International Transport Forum, Transport-related $CO_2$ emissions of the tourism sector: Modelling results. Report, 2019.

[9]   Sharpley, R., Tourism and sustainable development: Exploring the theoretical divide. *J. Sustain. Tour.*, **8**, pp. 1–19, 2000.

[10]  Pan, S.Y., Gao, M., Kim, H., Shah, K.J., Pei, S.L. & Chiang, P.C., Advances and challenges in sustainable tourism toward a green economy. *Sci. Total Environ.* **635**, pp. 452–469, 2018.

[11]  Aall, C., Sustainable tourism in practice: Promoting or perverting the quest for a sustainable development. *Sustain.*, **6**, pp. 2562–2583, 2014.

[12] Park, D.B. & Yoon, Y.S., Segmentation by motivation in rural tourism: A Korean case study. *Tour. Manag.,* **30**, pp. 99–108, 2009.

[13] Barget, E. & Gouguet, J.J., The impact on tourism of mega-sporting events: The stakes of foreign spectators. *Tourism Review International,* **16**(1), pp. 75–81, 2012.

[14] Michailidou, A.V., Vlachokostas, C., Moussiopoulos, N. & Maleka, D., Life cycle thinking used for assessing the environmental impacts of tourism activity for a Greek tourism destination. *J. Clean Prod.,* **111**, pp. 499–510, 2016.

[15] Maugeri, E., Gullo, E., Romano, P., Spedalieri, F. & Licciardello, A., The bioeconomy in Sicily: New green marketing strategies applied to the sustainable tourism sector. *Engineering and Management,* **4**(3), pp. 133–142, 2017.

[16] De Camillis, C., Raggi, A. & Petti, L., Life cycle assessment in the framework of sustainable tourism: A preliminary examination of its effectiveness and challenges. *Progress in Industrial Ecology,* 7(3), pp. 205–218, 2010.

[17] ISO 14067, Greenhouse gases – Carbon footprint of products – Requirements and guidelines for quantification and communication, 2013. https://www.iso.org/standard/59521.html. Accessed on: 15 Apr. 2022.

[18] Chan, E.S.W. & Hsu, C.H.C., Environmental management research in hospitality. *Int. J. Hosp. Manag.,* **28**(5), pp. 886–923, 2016.

[19] Eluwole, K.K., Akadiri, S.S., Alola, A.A. & Etokakpan, M.U., Does the interaction between growth determinants a drive for global environmental sustainability? Evidence from world top 10 pollutant emissions countries. *Sci. Tot. Environ.,* **705**, 135972, 2020.

[20] König, H., Schmidberger, E. & De Cristofaro, L., Life cycle assessment of a tourism resort with renewable materials and traditional construction techniques in Portugal. *Sustainable Construction, Materials and Practices: Challenge of the Industry for the New Millenium's Conference,* p. 8, 2007.

[21] Foolmaun, R.K. & Ramjeeawon, T., Disposal of post-consumer polyethylene terephthalate (PET) bottles: Comparison of five disposal alternatives in the small island state of Mauritius using a life cycle assessment tool. *Environ. Technol.,* **33**(5), pp. 563–572, 2012.

[22] ISO 14040, Environmental management – Life cycle assessment – Principles and framework, 2006. https://www.iso.org/standard/37456.html. Accessed on: 15 Apr. 2022.

[23] De Camillis, C., Raggi, A. & Petti, L., Tourism LCA: State-of-the-art and perspectives. *Int. J. Life. Cycle. Assess.,* **15**(2), pp. 148–155, 2010.

[24] ISO 14044, Environmental management – Life cycle assessment – Principles and framework, 2006. https://www.iso.org/standard/38498.html#:~:text=ISO%2014044%3A2006%20specifies%20requirements,and%20critical%20review%20of%20the. Accessed on: 18 Apr. 2022.

[25] Sharp, H., Grundius, J. & Heinonen, J., Carbon footprint of inbound tourism to Iceland: A consumption-based life-cycle assessment including direct and indirect emissions. *Sustainability,* **8**(11), p. 1147, 2016.

[26] Hu, A.H., Huang, C.Y., Chen, C.F., Kuo, C.H. & Hsu, C.W., Assessing carbon footprint in the life cycle of accommodation services: The case of an international tourist hotel. *Int. J. Sustain. Dev. World Ecol.,* **22**(4), pp. 313–323, 2015.

[27] Filimonau, V., Dickinson, J., Robbins, D. & Reddy, M.V., The role of 'indirect' greenhouse gas emissions in tourism: Assessing the hidden carbon impacts from a holiday package tour. *Transportation Research Part A: Policy and Practice,* **54**, pp. 78–91, 2013.

[28]  Lai, J.H.K., Carbon footprints of hotels: Analysis of three archetypes in Hong Kong. *Sustainable Cities and Society*, **14**, pp. 334–341, 2015.

[29]  Priyadarsini, R., Xuchao, W. & Eang, L.S., A study on energy performance of hotel buildings in Singapore. *Energy and Buildings*, **41**, pp. 1319–1324, 2009.

[30]  Luo, F., Becken, S. & Zhong, Y., Changing travel patterns in China and 'carbon footprint' implications for a domestic tourist destination. *Tour. Manag.*, **65**, pp. 1–13, 2018.

[31]  Michailidou, A.V., Vlachokostas, C., Achillas, C., Maleka, D., Moussiopoulos, N. & Feleki, E., Green tourism supply chain management based on life cycle impact assessment. *Environ. Sci. Eur.*, **6**(1), pp. 30–36, 2016.

[32]  GaBi, Introduction to LCA and modelling using GaBi: Part 1, 2021. https://gabi.sphera.com/international/support/gabi-learning-center/gabi-learning-center/part-1-lca-and-introduction-to-gabi/. Accessed on: 20 Apr. 2022.

[33]  Solé, A., Miró, L. & Cabeza, L.F., Environmental approach. *High-Temperature Thermal Storage Systems Using Phase Change Materials*, pp. 277–295, 2018.

[34]  Mancini, M.S., Galli, A., Niccolucci, V., Lin, D., Bastianoni, S., Wackernagel, M. & Marchettini, N., Ecological footprint: Refining the carbon footprint calculation. *Ecol. Indic.*, **6**, pp. 390–403, 2016.

[35]  Department for Environment, Food and Rural Affairs (DEFRA), A research report for the Department for Environment, Food and Rural Affairs by the Stockholm Environment Institute and the University of Minnesota. Report, 2008.

[36]  Michailidou, A.V., Vlachokostas, C. & Moussiopoulos, N., A methodology to assess the overall environmental pressure attributed to tourism areas: A combined approach for typical all-sized hotels in Chalkidiki, Greece. *Ecol. Indic.*, **50**, pp. 109–119, 2015.

[37]  Rosselló-Batle, B., Moià, A., Cladera, A. & Martínez, V., Energy use, $CO_2$ emissions and waste throughout the life cycle of a sample of hotels in the Balearic Islands. *Energy Build.*, **42**(4), pp. 547–558, 2010.

[38]  Gössling, S., Peeters, P., Ceron, J.P., Dubois, G., Patterson, T. & Richardson, R.B., The eco-efficiency of tourism. *Ecological Economics*, **54**(4), pp. 417–434, 2005.

[39]  Li, L., Li, J., Tang, L. & Wang, S., Balancing tourism's economic benefit and $CO_2$ emissions: An insight from input–output and tourism satellite account analysis. *Sustainability*, **11**(4), p. 1052, 2019.

[40]  UNWTO, Climate change and tourism. *Second International Conference on Climate Change and Tourism*. Report, 2008.

[41]  UNWTO, International tourism highlights, 2019. https://www.e-unwto.org/doi/pdf/10.18111/9789284421152. Accessed on: 20 Apr. 2022.

[42]  Sun, Y.Y., Decomposition of tourism greenhouse gas emissions: Revealing the dynamics between tourism economic growth, technological efficiency, and carbon emissions. *Tour. Manag.*, **55**, pp. 326–336, 2016.

[43]  Greiff, K., Teubler, J., Baedeker, C., Liedtke, C. & Rohn, H., Material and carbon footprint of household activities. Living Labs: Design and Assessment of Sustainable Living, eds D. Keyson, O. Guerra-Santin & D. Lockton, Springer: Cham, pp. 259–275, 2016.

[44]  El Hanandeh, A., Quantifying the carbon footprint of religious tourism: The case of Hajj. *J. Clean Prod.*, **52**, pp. 53–60, 2013.

[45]  Santana, M.V.E., Cornejo, P.K., Rodríguez-Roda, I., Buttiglieri, G. & Corominas, L., Holistic life cycle assessment of water reuse in a tourist-based community. *J. Clean Prod.*, **233**, pp. 743–752, 2019.

[46]  Scheepens, A.E., Vogtländer, J.G. & Brezet, J.C., Two LCA based methods to analyse and design complex (regional) circular economy systems. Case: making water tourism more sustainable. *J. Clean Prod.,* **114**, pp. 257–268, 2016.

[47]  Cerutti, A.K., Beccaro, G.L., Bruun, S., Donno, D., Bonvegna, L. & Bounous, G., Assessment methods for sustainable tourism declarations: The case of holiday farms. *J. Clean Prod.,* **111**, pp. 511–519, 2016.

[48]  Sun, R.H., Ye, X.L., Gao, J., Zhu, Z.F., Du, J., Study on carbon footprint and spatial distribution characteristics of human activities in Jiuzhai Valley scenic area. *Applied Ecology and Environmental Research,* **17**(4), pp. 7477–7493, 2019.

[49]  Gallucci, T. & Dimitrova, V., The role of carbon footprint indicator for sustainable implications in tourism industry: Case study of Bulgaria. *Int. J. Sustain. Dev. World Ecol.,* **12**(1), p. 61, 2020.

[50]  Cadarso, M., Gómez, N., López, L.A. & Tobarra, M., Calculating tourism's carbon footprint: Measuring the impact of investments. *J. Clean Prod.,* **111**, pp. 529-537, 2016.

[51]  Kitamura, Y., Ichisugi, Y., Karkour, S. & Itsubo, N., Carbon footprint evaluation based on tourist consumption toward sustainable tourism in Japan. *Sustainability,* **12**(6), pp. 1–23, 2020.

[52]  Baumber, A., Merson, J. & Lockhart, C., Promoting low-carbon tourism through adaptive regional certification. *Climate,* **9**(1), pp. 1–22, 2021.

[53]  Dorta, P., Díaz, J., López, A. & Bethencourt, C., Tourism, transport and climate change: The carbon footprint of international air traffic on islands. *Sustainability,* **13**(4), p. 1795, 2021.

[54]  Gühnemann, A., Kurzweil, A. & Mailer, M., Tourism mobility and climate change: A review of the situation in Austria. *J. Outdoor Recreat. Tour.,* 100382, 2021.

[55]  Kitamura, Y., Karkour, S., Ichisugi, Y. & Itsubo, N., Evaluation of the economic, environmental, and social impacts of the COVID-19 pandemic on the Japanese tourism industry. *Sustainability,* **12**(24), pp. 1–21, 2020.

# SUSTAINABLE PRACTICES IN HOTELS IN SRI LANKA: ANALYSIS OF ENVIRONMENTALLY SUSTAINABLE ASPECTS

N. W. THUSITHA DILSHAN & AYAKO TOKO
Faculty of International Tourism Management, Toyo University, Japan

## ABSTRACT

Sustainable practices are particularly important in the tourism industry, and a substantial number of studies have shown that different environmentally sustainable practices (hereinafter written as ESP) are used by hoteliers who want to be responsible in the industry. On the other hand, hotel practices have been changing according to the guests' expectations to attract more guests; thus, filling the gap between guests' expectations and satisfaction of ESP is important. Based on that, this study undertook quantitative research with aiming to identify the gap between guests' expectations and satisfaction of ESP in Sri Lankan hotels. The study adopted five dimensions related to ESP which were identified through a preliminary study of previous literature. An empirical survey was conducted online with a structured questionnaire to identify the guests' expectations. Results of factor analysis revealed that environmentally-friendly production, waste management, food and beverage production, and environmentally-friendly hotel construction are the priorities of guests' expectations. Moreover, waste management, biodiversity conservation and food and beverage production represent a significant gap between the guests' expectations and satisfaction of ESP. Finally, this study found that guests' expectations mainly rely on the superficial natural atmosphere; therefore, more appropriate information and interpretation of sustainability would be needed for guests to optimize the usage of natural resources. Moreover, the study recommends that determining the reasons for making guests unsatisfied and find the solutions to minimize the dissatisfaction are important since currently available ESP have not met the satisfaction level of the guests.
*Keywords: sustainable tourism, environmentally sustainable practices, hotel industry, guests' expectations, factor analysis, Sri Lanka.*

## 1 INTRODUCTION

Tourism has become a great contributor to the economy in many developing countries. For instance, it contributed 12.6% to the Sri Lankan GDP in 2019 becoming the third major foreign exchange earning sector [1]. This rapid development of the industry in the last few decades has made many opportunities for the stakeholders such as bringing foreign exchange and generating employment. At the same time, tourism makes a negative impact as well due to the huge market competition. Hence, sustainability was required to the tourism industry to mitigate the negative impacts and sustainable tourism has emerged as an alternative tourism product from the 1990s with the aim of minimizing the negative impact to all economic, social and environmental aspects in the tourism industry [2], [3].

The term "sustainable development" became famous worldwide through the publication of the "Brundtland Report" (Our Common Future) by the World Commission on Environment and Development (WCED) in 1987. Since then, sustainable development has been continuously discussed in the United Nations Conferences (Earth Summit in 1992, Millennium Summit in 2000, World Summit on Sustainable Development in 2002, Rio+20 in 2012, and Sustainable Development Summit in 2015) and SDGs (Sustainable Development Goals) are now in process as shown in Fig. 1.

WIT Transactions on Ecology and the Environment, Vol 256, © 2022 WIT Press
www.witpress.com, ISSN 1743-3541 (on-line)
doi:10.2495/ST220111

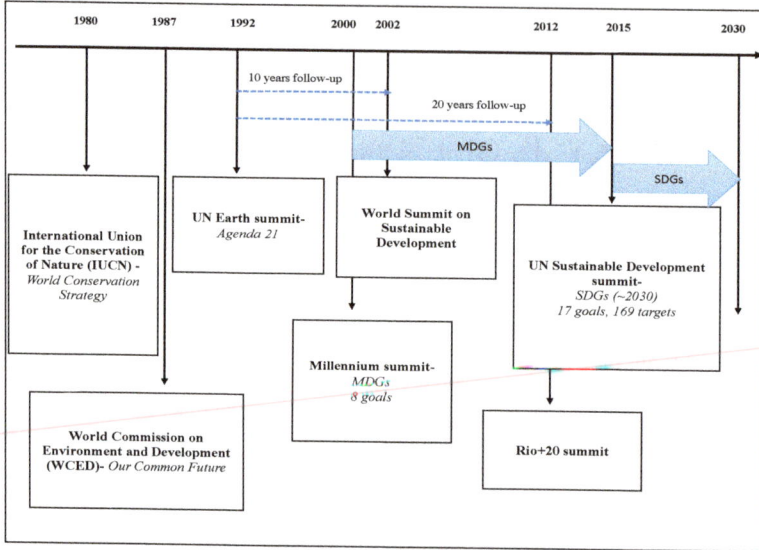

Figure 1:  Transition of arguments of sustainable development arguments. *(Source: Developed by the author referring to [4]–[6].)*

Accordingly, sustainable tourism also became popular in the late 1980s with the aim of increasing the positive effects of tourism on the environment and livelihoods of local communities [7]. In order to achieve sustainable tourism development, three basic elements need to be aligned as economic growth, social inclusion, and environmental protection. These elements are interrelated, and all are essential to the wellbeing of individuals and societies. Most commonly accepted definition of sustainable tourism refers to "an activity that takes full account of its current and future economic, social, and environmental impacts, addressing the needs of visitors, the industry, and the environment and host communities" [8]. Moreover, sustainable tourism was highlighted in the UN Sustainable Development Summit in 2015 because of three goals (it is expected that Goal 8 (Decent Work and Economic Growth), Goal 12 (Responsible Consumption and Production), and Goal 14 (Life Below Water) would be directly supported and promoted by tourism) among the 17 SDGs that are directly associated with the tourism industry, which are inclusive and sustainable economic growth, sustainable consumption and production, and the sustainable use of oceans and marine resources [9].

Hotels are a main sector in the tourism industry which consumes more energy during their operations and their negative impact on nature is also significantly high [3]. As a result, the concept of environmentally-friendly hotels appeared in developed countries in the 1980s [10]. Later, some hotels gradually started to be concerned about the environment and tried their best at changing strategic concepts, improving awareness of the environment and enforcing environment friendly management [10]. According to Akhtar and Najar [11], environmentally sustainable practices (hereinafter written as ESP) in the tourism industry under three major areas of energy management, waste management and water conservation has gained more importance. However, there are many new areas emerging recently in the hotel sector such as biodiversity conservation, environmental education, eco-friendly design, passive design and product innovation [12]. Correspondingly, considerable number of studies have focused on ESP in the hotels such as segregation of garbage, waste recycle, water conservation, plantation of trees, reuse of unused guests' amenities, paper free office, reuse of linen, electronic key card system, and reusable amenity bottles, by using different analytical methods as descriptive analysis, MANOVA, quality function deployment, factor analysis, and content analysis [11]–[15]. On the other hand, customers are one of key stakeholders of tourism industries and some of them try to demand ESP while paying additional money for sustainable practices and their expectations could be influenced by their knowledge of sustainability activities [16], [17]. Referring to the previous studies, different types of guests' expectations seem to be affected by such as hotel constructions without harming to the environment, atmosphere which harmonies with natural environment, presence of automatic electricity controllers, presence of low-impact interiors, reuse of towels and linens, effective use of water, green products, organic foods and so on [14], [16], [18]–[23].

Considering the current situation of tourism industry and related research explained above, it is significance to focus on ESP of hotels and guests' expectations. At the same time, the authors regard that environmentally sustainable management is a critical issue especially for hotels in developing countries that are vulnerable to environmental changes and want to gain more attractions of responsible tourists who could behave less harmful to the environment. Therefore, this study is accordingly focused on Sri Lanka which is one of leading countries of sustainable tourism in developing countries and already has national certification scheme for the hotel sector. As Ratnayake and Miththapala [24] mentioned that there is a significant potential to improve the sustainable practices in the hotel sector in the county, and another study also revealed that ESP in Sri Lankan hotels help to improve the competitiveness, save money, and attract environmentally concerned customers [25]. On the other hand, there are few studies that focused on guests' expectations that substantially affect hotels' benefits and sustainability. Thus, this study aims to (1) prioritize guests' expectations of ESP in Sri Lankan hotels, and (2) identify the gap between guests' expectations and satisfaction. Results are expected to develop a new method for similar research and to contribute to further improvement of ESP in hotel sector.

## 2 SUSTAINABLE TOURISM IN SRI LANKA

Sri Lankan tourism industry formally started its developments in 1966 with the establishments of Ceylon Tourist Board (Ceylon Tourist Board was enabled by the Ceylon Tourist Board Act No. 10 of 1966) and the Ceylon Hotels Corporation (Ceylon Hotels Corporation enabled by Ceylon Hotels Corporation Act of 1966) [26]. However, industry experienced a hard period during the thirty years of continued civil war. Again, tourism has become the fastest growing sector in the post-civil war era since 2009 and [27] an adequate number of hotel rooms is a major requirement when the number of tourists increases [28]. Thus, a number of national and international companies have started to invest in the hotel sector. Implementation of sustainable practices in hotels is a key driver towards tourism sustainability [11] and Sri Lankan hoteliers also implemented different sustainable practices in the properties. As a result, Sri Lankan Tourism Development Authority (SLTDA) also started a certification system called National Sustainable Tourism Certification (NSTC) from 2019 with the purpose of evaluating the sustainable practices in the hotels. It was the first time in the world a government tourism organization has taken such a kind of initiative. NSTC is the first step towards making the tourism industry sustainable with the technical and financial assistance of the Biodiversity Finance Initiative (BIOFIN) of the United Nations Development Program (UNDP). Moreover, the certification will be relevant internationally since the global context was considered in designing the project with the facilitation of the Global Sustainable Tourism Council (GSTC) [29]. Thirty-seven hotels have been recognized for awards on impressive sustainable practices in different categories through an internal and external hotel auditing process under platinum, gold, silver and bronze categories. In addition, four special categories were also recognized as the hotels which had the best specialized practices.

According to Wijesundara [30], Sri Lankan hoteliers have heavily incorporated with economic and social aspects while environmental concerns were in questionable level. Furthermore, Akhtar and Najar [11] also mentioned that environmental sustainability is gaining more importance out of the three dimensions of sustainability and many hotels have already started to implement ESP in their daily business routine. In addition, Abdou et al. [31] also highlighted that, if the resources are not managed properly, they will negatively impact on the environment and operating costs of an organization. Consequently, it brings a significant importance to set priorities guests' expectations and fill the gap between guests' expectations and satisfaction in the ESP in Sri Lankan hotels.

## 3  METHODS

According to the objectives of (1) prioritize guests' expectations of ESP in Sri Lankan hotels; and (2) identify the gap between guests' expectations and satisfaction of ESP in Sri Lankan hotels, interdisciplinary methods have been applied in the study, and it can be divided into two approaches: (i) literature review and analysis to develop a structured questionnaire, and (ii) empirical survey with a developed questionnaire to identify the guests' expectations and satisfaction. The research conducted online survey because of the limitation of COVID-19 pandemic, and data was collected by using Google form questionnaires. The research targeted two samples as inbound travellers and domestic travellers. The snowball sampling technique was used to collect data and questionnaires was designed in the English language. Google forms were sent through direct emails and social media during the period from December 2020 to April 2021. Totally, 257 respondents' data were collected, and 249 among them was reliable enough for the analysis, which included descriptive analysis and the Exploratory Factor Analysis (EFA) according to the literature review [10], [18], [24]. The detailed process is explained in Table 1.

Table 1:  Materials and methods according to the research objectives.

| Research objectives | Material and methods |
| --- | --- |
| Prioritize guests' expectations of ESP in Sri Lankan hotels | Literature, secondary data, structured questionnaires (online) and Exploratory Factor Analysis (EFA) |
| Identify the gap between guests' expectations and satisfaction of ESP in Sri Lankan hotels | Literature, structured questionnaires (online) and quantitative analysis |

First, the literature review was conducted by the authors to capture the baseline information related to the environmentally sustainable tourism and sustainable practices of Sri Lankan hotels. In addition to various articles from international academic journals, further documents, books, secondary data including UNWTO reports, hotels' annual reports and sustainable policies in the hotels were collected during the literature review process. Subsequently, structured questionnaire was developed based on the results of preliminary research of literature review with a simple form by mainly including closed-ended and open-ended questions as shown in Table 2, which was then used to collect data from the inbound and domestic tourists who have visited Sri Lankan sustainable hotels before the COVID-19 pandemic. As shown in Table 2, questionnaires consist of three main sections. Questions in the first section are used to identify the demographic features of the respondents and the other two sections were developed to fulfil the main research objectives of the study. Five Likert scales were used to measure the expectation level and the satisfaction level in both section 2 and section 3.

Resulting from the literature review, section 2 and section 3 in Table 2, are consisted of five dimensions. Then, as shown in Table 3, each dimension is coded in a specific way with having four indicators.

Table 2: Categories of the questionnaire.

| Section 1: | Demographic questions: | Gender, age, region |
|---|---|---|
| Section 2: | Expectations regarding sustainable practices: | Willingness to pay additional fee on sustainable practices; Expectations from a sustainable hotel |
| Section 3: | Sustainable practices in the hotels: | Satisfaction level of the ESP in Sri Lankan hotels |

Table 3: Dimensions and indicators. *(Source: Developed by the author referring to [11]–[16], [18]–[25].)*

| Dimensions | Indicator cording | Indicators |
|---|---|---|
| Biodiversity conservation (BC) | BC1 | The hotel construction not causing any harm to the environment |
| | BC2 | Conserving the natural habitats surrounding the hotel |
| | BC3 | Staff having a good knowledge of environmental conservation |
| | BC4 | Having environmental education programs for visitors |
| Environmentally friendly energy management (EM) | EM1 | Using natural lighting and ventilation |
| | EM2 | Electricity generates from alternative resources (Solar, biomass, etc.) |
| | EM3 | Availability of automated electricity controllers (Automatic switch system, key cards, etc.) |
| | EM4 | Availability of electric buggy carts for transportation |
| Solid waste and water management (WM) | WM1 | Separation of solid waste by rubbish bins |
| | WM2 | Reuse and recycle the solid waste |
| | WM3 | Availability of re-usable drinking water bottles |
| | WM4 | Treatment and reuse of the wastewater |
| Food and beverage production (FBP) | FBP1 | Producing vegetables and fruits from own garden |
| | FBP2 | Buying organic products from surrounding |
| | FBP3 | Preparing foods using organic vegetables |
| | FBP4 | Providing fresh fruit juices |
| Construction and production through environmentally friendly (EFC) | EFC1 | Hotel construction from the natural materials |
| | EFC2 | Room interior designs from the natural materials |
| | EFC3 | Amenities made from natural or reusable materials |
| | EFC4 | Gifts and souvenirs made from natural or reusable materials |

Official Facebook pages and Instagram accounts of the hotels and travel agencies were used to identify the travellers who visited hotels in Sri Lanka. Google form questionnaires were distributed on social media platform such as Facebook personal accounts, groups and Instagram accounts. Additionally, a link of the Google form was sent to the email address collected from the Instagram users who have commented and tagged photos related to sustainable hotels in Sri Lanka. Furthermore, TripAdvisor users who have commented and tagged photos about sustainable hotels were another online source used to collect data for the survey. Totally, the link has been shared among 1200 travellers and the response rate is 21.4%. Finally, 249 respondents' data was reliable enough to use for the analysis.

## 4  PILOT SURVEY

The developed questionnaire was tested by the pilot survey to measure the reliability. According to the structure of the research, there are two sections included to check Cronbach's alpha value. Thus, preliminary analysis has measured the reliability of the guests' expectations and satisfaction level with sustainable hotel practices by using Statistical Package for the Social Sciences (SPSS) software. According to Smith et al. [34], the requirement for a reliable questionnaire is that Cronbach's alpha value must exceed 0.7. The pilot survey selected 20 respondents' samples to ensure meeting the requirement. According to the results, both Cronbach's alpha value belong to the guests' expectations ($\alpha 1=0.90$) and satisfaction($\alpha 2=0.89$) of the ESP in the hotels have exceeded 0.7. It proves that the questionnaire was reliable enough to collect the data for the survey since $\alpha 1 > \alpha 2 > 0.7$.

## 5  RESULTS

### 5.1  Demographic features

According to the empirical survey, 123 inbound traveller and 126 domestic traveller responses have been collected during the data collection period. As per the results as shown in Table 4, There is no big difference between male and female respondents. The highest respondents age category is 30–39 and the majority of the inbound guest respondents are from the Asia and Pacific region.

Table 4:  Demographic information of the respondents (%).

| | | Inbound (n = 123) | Domestic (n = 126) | Total (n = 249) |
|---|---|---|---|---|
| Region | Domestic | 50.60% | | |
| | Africa | 1.20% | | |
| | America | 4.02% | | |
| | Asia and Pacific | 22.89% | | |
| | Europe | 17.27% | | |
| | Middle East | 4.02% | | |
| Gender | Male | 51.22% | 49.21% | 50.20% |
| | Female | 48.78% | 50.79% | 49.80% |
| | | Inbound % | Domestic% | Total% |
| Age group | 19 or below | 0.81% | 1.59% | 1.20% |
| | 20–29 | 11.38% | 39.68% | 25.70% |
| | 30–39 | 48.78% | 45.24% | 46.99% |
| | 40–49 | 24.39% | 7.14% | 15.66% |
| | 50–59 | 14.63% | 3.97% | 9.24% |
| | 60 and over | 0.00% | 2.38% | 1.20% |

## 5.2 Willingness to pay additional fee on sustainability practices

According to the results, 74% of the respondents are willing to pay an additional fee on sustainable practices as shown in Fig. 2, and there is no big difference between inbound and domestic tourists. "How much" was not asked in this survey because the price level is varied in different hotel classes and it is not an objective of this question.

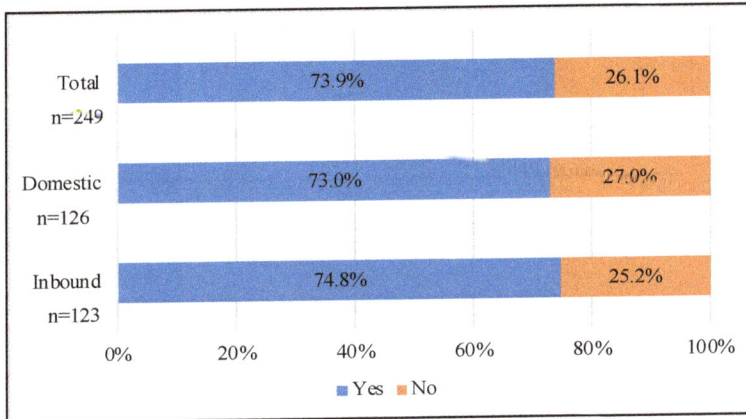

Figure 2:  Willingness to pay an additional fee on sustainable practices.

## 5.3 Guests' expectations in Sustainable hotels in Sri Lanka

For the analysis of the objective (1), the study used EFA to prioritize the guests' expectations. When proceeding a EFA it is necessary that the sample should be greater than 0.5 for Kaiser–Meyer–Olkin (KMO) value and less than 0.005 for Bartlett's test significant value. KMO and Bartlett's test measure the strength of the relationship among variables. As shown in Table 5, it illustrates that the KMO value is 0.864 and Bartlett's test significant value is 0.000. Therefore, both KMO value and Bartlett's test significant value fulfil the requirements. Consequently, the sample is adequate for the EFA to prioritize the guests' expectations of ESP in Sri Lankan hotels.

Table 5:  KMO and Bartlett's test.

| Kaiser–Meyer–Olkin (KMO) Measure of Sampling Adequacy | | 0.864 |
|---|---|---|
| Bartlett's Test of Sphericity | Approx. Chi-Square | 2268.51 |
| | df | 190 |
| | Sig. | 0.000 |

According to Scree Plot in Fig. 3 which displays the distribution of eigenvalue, there are five components exceeding the eigenvalue of 1. These components are useful for the EFA.

Among the different types of expectations which were examined by indicators explained in Table 3, (I) environmentally friendly production and waste management, (II) food and beverage production, (III) environmentally friendly hotel construction, (IV) natural habitats

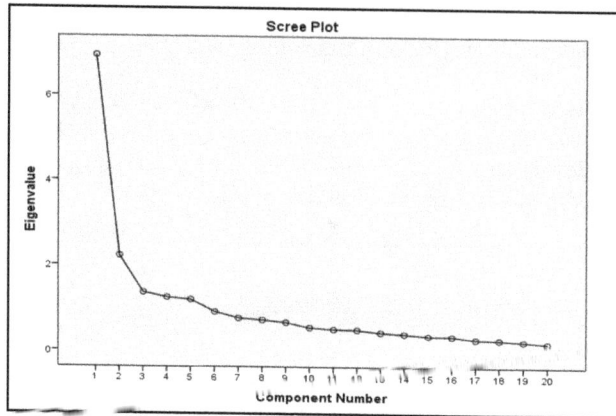

Figure 3:  Scree plot.

conservation, and (V) visitor programs about environmental protection have the higher eigenvalue in the order as shown in Table 6. Additionally, indicators belong to each expectation dimension also as shown in Table 6. Furthermore, these five components represent the 64.73% of the total variances among all components.

Table 6:  Priorities of guests' expectations in Sri Lankan hotels.

| Statement | Number of indicators | Eigen value | Variance % | Indicators |
|---|---|---|---|---|
| (I)   Environmentally friendly production and waste management | 13 | 6.94 | 34.70% | BC3, EM1, EM2, EM3, EM4, WM1, WM2, WM3, WM4, EFC1, EFC2, EFC3, EFC4 |
| (II)  Food and beverage production | 4 | 2.22 | 11.08% | FBP1, FBP2, FBP3, FBP4, FBP4 |
| (III) Environmentally friendly hotel construction | 1 | 1.36 | 6.78% | BC1 |
| (IV) Natural habitats conservation | 1 | 1.22 | 6.22% | BC2 |
| (V)  Visitor programs about environmental protection | 1 | 1.19 | 5.95% | BC4 |

Moreover, analysis identified the below ESP with higher factor loading, which were the highly expected practices of the guests in Sri Lankan hotels.

Table 7:  Highly expected ESP.

| Indicators | | Factor loading |
|---|---|---|
| Room interior designs from the natural materials | EFC2 | 0.751 |
| Amenities made from natural or reusable materials | EFC3 | 0.723 |
| Gifts and souvenirs made from natural or reusable materials | EFC4 | 0.715 |
| Treat and reuse the wastewater | WM4 | 0.658 |
| Hotel constructions from the natural materials | EFC1 | 0.655 |

### 5.4  Gap between guests' expectations and satisfactions on ESP

Empirical study has examined and identified the gap between the guests' expectations and satisfaction about ESP by using the five Likert scale. As per the results of the study as shown in Fig. 4, solid waste and water management (WM), biodiversity conservation (BC) and food and beverage production (FBP) are the dimensions which have significant gaps.

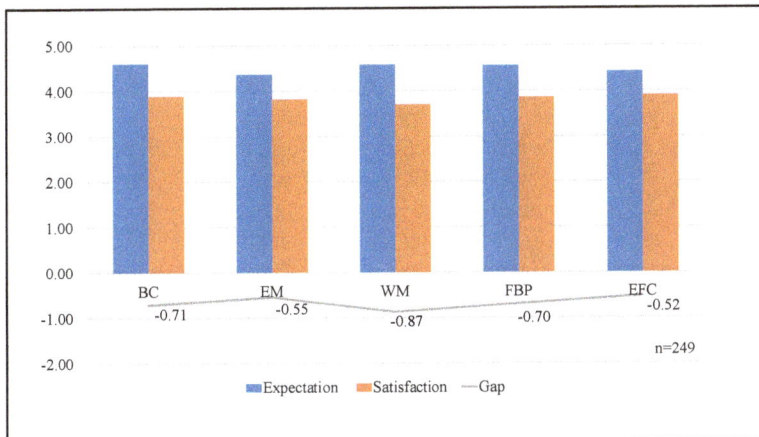

Figure 4:  Gap between guests' expectations and satisfactions on ESP.

In addition, the study determined the gaps between guests' expectations and satisfactions of the ESP in practice wise by using eqn (1).

$$Sj = \frac{\sum_{i=1}^{n}(SATi-EXPi)}{\sum_{i=1}^{n}(EXPi)} \ X \ 100\% \tag{1}$$

where S = satisfaction gap; SAT = satisfaction value; EXP = Expectation value; $i$ = number of respondents; $j$ = sustainable practice.

Based on eqn (1), percentages of the gaps have been calculated for all twenty indicators. Consequently, the below mentioned five practices have the higher gaps among the indicators. The significance of the result is that all practices have a negative value (–) and the negative values indicate that the expectations are higher than the satisfaction.

1. Having environmental education programs for visitors   (BC4)   –22.7%
2. Separation of solid waste by trash bins   (WM1)   –20.4%

3. Buying organic products from surrounding     (FBP2)    −19.9%
4. Reusing and recycling the solid waste     (WM2)    −19.3%
5. Availability of re-usable drinking water bottles     (WM3)    −19.3%

## 6 DISCUSSION

According to the results of the study, five dimensions for ESP in Sri Lankan hotels have been firstly identified through the literature review analysis. In addition, each dimension consisted of four ESP, which were used as indicators in this study. The results of the empirical survey proved that majority of the respondents were willing to pay additional fee on sustainable practices, which has been also revealed by the previous research [14], [16], [19]. Furthermore, factor analysis emphasized that there were two critical areas in the prioritization of the guests' expectation in Sri Lankan hotels. The highest prioritized area was environmentally friendly production and waste management consisting of thirteen indicators, and the second highest area was food and beverage production consisting of four indicators. Furthermore, highly expected practices were centered in environment friendly construction (EFC); i.e., EFC2, EFC3, and EFC4. It indicates that guests tend to have higher expectations with practices that are directly observable and affect their comfortable stay such as room interiors, hotel constructions, souvenirs and amenities. Results explained above presented the priority areas and practices of guests' expectations, which would be useful to make a management strategy to meet guests' expectations. In addition, according to previous studies, ensuring the guests' satisfaction is required in the hotel operation of Sri Lanka especially in ESP [25] and it creates many benefits to the hotels as well [24], [25]. Thus, this study clarified the significant gaps between guests' expectations and satisfaction on both dimensions and indicators. Among the dimensions, WM, BC and FBP have the higher gaps, and regarding the indicators, higher gaps are mainly found in biodiversity conservation, waste management and food and beverage production. Moreover, environmental education programs, separation of solid waste by using trash bins inside the premises, buying organic products from hotel surrounding, reusing and recycling the collected wastage, and providing glass water bottles instead of the plastic ones are detailed ESP where gaps are significantly highlighted. The findings and applied methods could help hotel management to mitigate the level of dissatisfaction of their guests towards ESP. Hence, this study identified the areas where more attention should be paid by the hotel management to develop the substantial number of ESP with fulfilling the guests' expectations in Sri Lankan hotels. It could contribute to improve the quality of the ESP from the perspectives of the guests; and ultimately, best practices with economic benefits can be promoted while meeting guests' satisfaction.

## 7 CONCLUSION AND RECOMMENDATIONS

According to the study, tourists tend to expect superficial aspects of ESP such as room interior, amenities, and gifts and souvenirs from natural materials. However, natural materials also bring another cause of over-use of natural resources or degradation of the natural environment. Thus, more information and interpretation of sustainability, for example, sustainably sourced not just natural, would be needed for guests in sustainable hotels to optimize the usage of natural resources. In addition, this study identified that guests were not wholly satisfied with currently available ESP in Sri Lankan hotels. Hence, revealing the reasons for dissatisfaction are important to create satisfied guests in the hotels. Furthermore, identify the hotel's management's perception of ESP is also important to initiate proper and sufficient number of practices which can be met the guests' expectations. Accordingly increasing number of satisfied guests will increase the market share and profits for the businesses.

This study only concerned about the hotel sector. However, considering the future research, there are many other emerging sectors in tourism industry such as travel and event sectors which are directly involving in sustainable activities. Therefore, this study suggests to research on such emerging sectors since guests' expectations are different from the hotel sector. Moreover, identifying the detailed amount of willingness to pay the additional fee on sustainable practices will also be significantly important for the future research. This research has been focused mainly on the pre-COVID-19 conditions. However, guests' expectations and hotel practices would be changed after the pandemic; for example, more concerns would be based on medical and chemical waste management in the hotels. Thus, research on the post-COVID-19 conditions is another suggestion from this study.

## REFERENCES

[1]  SLTDA, *Annual Statistical Report: 2019*, Sri Lanka Tourism Development Authority, 2019.

[2]  Ceylan, E.N. & Tülbentçi, T., Example of an ecotourism farm in the context of sustainability: Pastoral valley ecological life farm. *International Journal of Advanced and Applied Sciences*, **7**(6), pp. 116–132, 2020.

[3]  Cevirgen, A., Baltaci, F. & Oku, O., Residents' perceptions towards sustainable tourism development: The case of Alanya. *International Symposium on Sustainable Development,* pp. 65–76, 2012.

[4]  IUCN, World Conservation Strategy, Conservation of Nature and Natural Resources, 1980.

[5]  WCED, *Our Common Future, World Commission on Environment and Development*, 1984.

[6]  United Nations, Agenda 21. https://sustainabledevelopment.un.org/outcomedocuments/agenda21. Accessed on: 28 Oct. 2021.

[7]  Smith, V.L. & Brent, M., *Hosts and Guests Revisited: Tourism Issues of the 21st Century*, pp. 28–41, 2001.

[8]  UNWTO, *Sustainable Tourism for Development Guidebook*, World Tourism Organization (UNWTO): Madrid, 2013.

[9]  UNWTO, Glossary of tourism terms. https://www.unwto.org/glossary-tourism-terms. Accessed on: 15 Oct. 2021.

[10]  Shen, H. & Zheng, L., Environmental management and sustainable development in the hotel industry: A case study from China. *International Journal of Environment and Sustainable Development,* **9**(1), pp. 194–127, 2010.

[11]  Akhtar, S. & Najar, A.H., Environmental sustainable practices in the hotels:from existence to implementation. *Ecology, Environment and Conservation*, **26**(1), pp. 111–116, 2020.

[12]  Jamaludina, M. & Yusof, Z.B., Best practice of green island resorts. *Social and Behavioral Sciences*, **105**, pp. 20–29, 2013.

[13]  Weaver, D., Davidson, M.C.G., Lawton. L., Patiar, A., Reid, S. & Johnston, N., Awarding sustainable Asia–Pacific hotel practices: Rewarding innovative practices or open rhetoric? *Tourism Recreation Research*, **38**(1), pp. 15–28, 2015.

[14]  Gilmore, E., Fuller, D. & Jo, J.H., Implementing sustainable business practices in a hotel and expanding green certification markets in the midwestern United States. *International Conference on Sustainable Building Asia*, pp. 270–274, 2014.

[15]  Eggeling, J., Sustainable tourism practices in the hospitality sector: A case study of Scandic, pp. 1–41, 2010.

[16] Mylan, J.A., Sustainable tourism in Costa Rica: Aligning tourists' interests with local development. *PURE Insights*, **7**(1), 2018.

[17] Sewwandi, A.M., Impact of sustainable practices on customer satisfaction in hotel industry: Evidence from boutique hotel in Southern Province in Sri Lanka. http://www.erepo.lib.uwu.ac.lk/handle/123456789/1315. Accessed on: 20 Dec. 2021.

[18] Dan-Cristian, D. & Raluca, B., An approach to sustainable development from tourists' perspective: Empirical evidence in Romania. *Business and Sustainable Development,* **15**(7), pp. 617–633, 2013.

[19] Agarwal, S. & Kasliwal, N., Going green: A study on consumer perception and willingness to pay towards green attributes of hotels. *International Journal of Emerging Research*, **6**(10), pp. 16–28, 2017.

[20] Andereck, K.L., Tourists' perceptions of environmentally responsible innovations. *Journal of Sustainable Tourism*, **17**(4), pp. 489–499, 2009.

[21] Berezan, O., Millar, M. & Raab, C., Sustainable hotel practices and guest satisfaction. *International Journal of Hospitality and Tourism*, **15**(1), pp. 1–10, 2014.

[22] Pakdil, F. & Kurtulmuşoğlu, F.B., Using quality function deployment for environmentally sustainable hotels: A combined analysis of customer and manager point of view. *European Journal of Tourism Research*, **16**, pp. 252–275, 2017.

[23] Ogbeide, G.C.A., Perception of green hotels in the 21st century. *Journal of Tourism*, **3**(1), pp. 1–17, 2012.

[24] Ratnayake, N. & Miththapala, S., A study on sustainable consumption practices in Sri Lanka hotel sector. *Civil Engineering Research for Industry,* pp. 89–94, 2011.

[25] Kularatne, T., Wilsona, C., Månsson, J., Hoang, V. & Lee, B., Do environmentally sustainable practices make hotels more efficient? A study of major hotels in Sri Lanka. *Tourism Management*, **71**, pp. 213–225, 2019.

[26] SLTDA, History. https://sltda.gov.lk/en/history. Accessed on: 18 Nov.2021.

[27] Kularatne, T., Wilsona, C., Månsson, J., Hoang, V. & Lee, B., Do environmentally sustainable practices make hotels more efficient? A study of major hotels in Sri Lanka. *Tourism Management*, **71**, pp. 213–225, 2019.

[28] Wij, I., *Sri Lanka Tourism: Poised for Growth*, HVS India: Gurgaon, 2011.

[29] UNDP, 37 hotels to be recognised for sustainable practices by SLTDA. https://www.lk.undp.org/content/srilanka/en/home/presscenter/pressreleases/2019/08/Hotels_to_be_recognised_for_sustainable_practices.html. Accessed on: 30 Oct. 2021.

[30] Wijesundara, C.N., Adoption of sustainable tourism practices by hotel operators in deep south of Sri Lanka. *International Journal of Research in Management Science and Technology*, **5**(6), pp. 5621–5630, 2017.

[31] Abdou, A.H., Hassan, T.H. & El Dief, M.M., A description of green hotel practices and their role in achieving sustainable development. *Sustainability*, **12**(22), 2020.

[32] Dabija, D.C. & Băbuţ, R., An approach to sustainable development from tourists' perspective. Empirical evidence in Romania. *Amfiteatru Economic Journal*, **15**(7), pp. 617–633, 2013.

[33] Mohamed, J. & Demerdash, E., Millennials' Viewpoints about Sustainable Hotels' Practices in Egypt: Promoting Responsible Consumerism. *International Journal of Humanities and Social Sciences,* **13**(5), pp. 622–629, 2019.

[34] Smith, J., Garratt, A., Guest, M., Greenhalgh, R. & Davies, A., Evaluating and improving health-relatedquality of life in patients with varicose veins. *Journal of Vascular Surgery*, **30**(4), pp. 710–719, 1999.

# Author index

**WIT**PRESS ...for scientists by scientists

# City Sustainability and Regeneration

*Edited by:* **S. MAMBRETTI**, *Polytechnic of Milan, Italy and* **J. L. MIRALLES I GARCIA,** *Politechnic University of Valencia, Spain*

A set of new studies are included in this volume which provides solutions that lead towards sustainability. Contributions originate from a diverse range of researchers, resulting in a variety of topics and experiences.

Urban areas face a number of challenges related to reducing pollution, improving main transportation and infrastructure systems and these challenges can contribute to the development of social and economic imbalances and require the development of new solutions. The challenge is to manage human activities, pursuing welfare and prosperity in the urban environment, whilst considering the relationships between the parts and their connections with the living world. The dynamics of its networks (flows of energy matter, people, goods, information and other resources) are fundamental for an understanding of the evolving nature of today's cities.

Large cities represent a productive ground for architects, engineers, city planners, social and political scientists able to conceive new ideas and time them according to technological advances and human requirements. The multidisciplinary components of urban planning, the challenges presented by the increasing size of cities, the amount of resources required and the complexity of modern society are all addressed.

ISBN: 978-1-78466-415-2    eISBN: 978-1-78466-416-9
Published 2020 / 164pp

*All prices correct at time of going to press but subject to change.*
*WIT Press books are available through your bookseller or direct from the publisher.*

www.ingramcontent.com/pod-product-compliance
Lightning Source LLC
Chambersburg PA
CBHW062008190326
41458CB00009B/3012